EXPERIMENTS IN CMOS TECHNOLOGY

Other Books in the
Advanced Technology Series

Experiments with EPROMS
By Dave Prochnow

Edited by Lisa A. Doyle

Experiments in
Artificial Neural Networks
By Ed Reitman

Edited by David Gauthier

Experiments in
Gallium Arsenide Technology
By D. J. Branning and Dave Prochnow

Edited by Lisa A. Doyle

EXPERIMENTS IN
CMOS TECHNOLOGY

Dave Prochnow
and
D.J. Branning

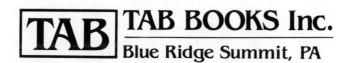

TAB BOOKS Inc.
Blue Ridge Summit, PA

FIRST EDITION
FIRST PRINTING

Copyright © 1988 by TAB BOOKS Inc.
Printed in the United States of America

Library of Congress Cataloging in Publication Data

Prochnow, Dave.
Experiments in CMOS Technology / by Dave Prochnow and D.J.
Branning.
p.   cm.
Bibliography: p.
Includes index.
ISBN 0-8306-9262-2   ISBN 0-8306-9362-9 (pbk.)
1. Metal oxide semiconductors, Complementary. 2. Metal oxide
semiconductors, Complementary—Experiments. I. Branning, D. J.
II. Title.
TK7871.99.M44P76 1988
621.3815′2—dc19   88-17066
                                                          CIP

TAB BOOKS Inc. offers software for
sale. For information and a catalog,
please contact TAB Software Department,
Blue Ridge Summit, PA 17294-0850.

Questions regarding the content of this book
should be addressed to:

Reader Inquiry Branch
TAB BOOKS Inc.
Blue Ridge Summit, PA 17294-0214

To my considerate and thoughtful Kathy

Trademark List

The following trademarked products are mentioned in *Experiments in CMOS Technology*.

Apple Computer, Inc.: Apple *IIe*
                          Macintosh
Autodesk, Inc.: AutoCAD 2
Bishop Graphics, Inc: E-Z Circuit
                          Quik Circuit
Borland International, Inc.: Turbo BASIC
                          Turbo C
Kepro Circuit Systems, Inc.: Kepro PCB
Ungar Electric: System 9000
Wahl Clipper Corporation: IsoTip 7800
                          IsoTip 7700
                          IsoTip 7240
Weller: EC2000
Wintek Corporation: HiWIRE
                          smARTWORK

Contents

Acknowledgments

Numerous contributions were made by six manufacturers during the preparation of this book. Autodesk, Inc., Bishop Graphics, Inc., Borland International, Heath Company, Kepro Circuit Systems, Inc., and Wintek Corporation each made generous hardware and/or software contributions which served as vital references for developing this text and its associated projects.

Introduction

You're probably wondering, just what's so advanced about CMOS technology? Essentially, CMOS fabrication techniques have been used in the construction of semiconductor devices for approximately thirty years. So, what *is* so advanced about CMOS technology?

Today's CMOS devices are stressing their benefits while minimizing their weaknesses. These new performance specifications read like a circuit-designer's wish list:

✤ A high input impedance.
✤ Minimal current drain.
✤ Enhancement of noise immunity.
✤ Reduction in vulnerability to static electrical discharge.

One emerging technology that is fully exploiting these newfound CMOS advancements is the biCMOS array. Roughly speaking, a biCMOS array is formed by the successful marriage of bipolar and CMOS circuits within the same *VLSI* (Very Large-Scale Integration) chip. This biCMOS union, for example, can result in the formation of a two-input NAND gate from six MOS transistors coupled to a two-transistor bipolar current booster. In this arrangement, current drain is reduced by selectively deactivating one of the bipolar transistors during static operation.

Unfortunately, to date, biCMOS technology has been restricted to the narrow confines of SRAMs, gate arrays, and composite analog/digital products. This restriction might ease as the biCMOS process is applied to VLSI RAMs, specifically dynamic RAMs (DRAMs).

One company that is devoting further research to the biCMOS issue is Texas Instruments. At TI, biCMOS technology is being combined with TI's LinCMOS (Linear CMOS) process for the creation of a hybrid biCMOS technology known as LinBiCMOS. This biCMOS enhancement combines high-voltage bipolar transistors with analog devices. The result is a mixed analog/digital package that can operate from a 20 V power supply at high speeds with minimal noise. These two attributes can be easily exploited in areas such as data communications, where driver/receiver circuitry is currently stressed by voltage and speed factors.

Another rising technological advancement in CMOS technology is in the fabrication of super-fast dynamic RAM (DRAM) chips. By using precharging techniques and NAND gate decoder/drivers, manufacturers such as Alliance Semiconductor Corporation have been able to deliver 60 ns access time, 1M-bit DRAMs.

In Alliance's case, the DRAM, a AS4C1002, is an 18-pin DIP 1M-by-1-bit package with a 100 ns cycle time. This design was made possible only through a major reduction in power supply noise. In order to minimize this speed-robbing CMOS attribute, two special circuit-level features were designed into the AS4C1002.

❖ A unique one-half power supply voltage precharge is used for supplying the needed charging and discharging currents. This process replaces the standard special false wordline pairs with a sequential procedure. Therefore, three pulses are used for discharging capacitances versus the standard single pulse method.

❖ The other area Alliance reduced noise in was selective decoder charging. Again, this technique bucks the current trend in decoder charging and discharging. In the AS4C1002, a NAND decoder/driver is used for restricting the required charging and discharging steps to a selected decoder. Therefore, the noise that is usually associated with multiple decoder charging/discharging cycles is reduced. One fringe benefit from this NAND decoder/driver configuration is an enhancement of circuit speed, through the elimination of a voltage drop between system ground and the power supply. The net result from these design innovations is a DRAM that operates at static RAM speeds.

In an effort to adequately examine these and other advances in CMOS technology, this book studies all of the significant CMOS devices that are currently available. Additional support is provided with a complete listing of each device's unique performance specifications. Furthermore, a detailed figure labels all of the pin assignments for each CMOS device. In short, this book is the definitive compendium of contemporary CMOS data. There is more to CMOS technology than just raw data sheets, however.

At the conclusion of each chapter, one or more complete projects will help you to experiment with today's CMOS devices. While some of these projects are complex in their construction, each one serves as an illustration of applying CMOS technology in a practical situation.

Several thorough appendices conclude this examination of CMOS technology. Project building suggestions, IC (Integrated Circuit) data sheets, manufacturers' addresses, a bibliography, and a glossary are contained within this concluding reference section.

Even though the use of CMOS fabrication techniques is a time-honored process, it is far from ancient history. Remember that semiconductor construction advances are currently being made that will make a firm understanding in CMOS technology vital for the future circuit designer.

1

CMOS Technology

Semiconductors have their roots (or gates) linked to a team of Bell Telephone Laboratory scientists. John Bardeen, Walter Brattain, and William Shockley won the 1956 Nobel Prize for physics for their landmark semiconductor research work begun in 1948. They named their semiconductor device the transistor.

Derived from the apparent function of this amplifier, the name transistor was the combination of TRANsfer and reSISTOR. These first transistors were constructed from two wire springs affixed to a semiconductor substrate. In designing with this construction technique, transistors made in this manner were called "point-contact" devices.

Manufacturing the semiconductor material was almost as difficult as joining the wire springs to the substrate. The most frequently used semiconductor materials in these early transistors were germanium and silicon. Both of these materials are Group IV semiconductor elements with four valence electrons, apiece. Production impurities in hydrogen-heated germanium dioxide necessitated the zone refining of each elemental form before fabrication into manufacturing wafers.

Following closely on the heels (or drain) of the point-contact transistor, was the junction transistor. Generally speaking, these junction transistors were divided into *PNP* and *NPN* devices. Beginning with the PNP transistor, two PN diodes are joined together with their polarities reversed. In other words, one of the diodes is forward biased, while the other one is reverse biased; a PNNP or PNP sandwich is formed. This arrangement leads to the flow of holes from the P to the N regions of the transistor. Only a small number of these holes recombine with electrons in the N region. The remainder of the holes diffuse through the base and travel to the collector of the transistor.

In contrast to the PNP transistor, the NPN transistor fixes a small P band in between two N regions. Therefore, the current flow in this transistor is in the form of electrons. This electron mobility results in an increased operation frequency for the NPN transistor over the PNP transistor.

DIGITAL LOGIC

Any discussion of digital electronics must begin with a solid introduction into digital logic. By definition, digital logic is the sequence of events or order that things occur within a digital circuit. This sequencing is governed by a strict application of mathematics. Entering into this mathematical logic scheme are the two possible conditions or states that can be present in a digital circuit: OFF or ON.

Several different names are given to this dual state depending on their circumstances of occurrence. For example, the two conditions of a digital signal are labeled as low and high for the OFF and ON states, respectively. Alternatively, in a graphics system, like a computer video display, the digital representation used for indicating the OFF and ON status of the individual pixels found on a monitor screen is either dark or light, respectively. Finally, in a microcomputer's *MPU* (Microprocessing Unit) these OFF and ON conditions are interpreted numerically as a 0 and a 1. This final symbolic definition will be used throughout the ensuing discussions of digital logic and introduction to digital memory.

These 0's and 1's of digital logic are manipulated with the *base-2* or *binary* number system. Like other number systems, the selective combination of the binary number system's 0 and 1 can be used for expressing any numeric value. Table 1-1 compares the same sequence of values for four different number systems. One unfortunate side effect to writing numbers in binary notation is their unwieldy dimensions. For example, consider the following decimal value along with its binary equivalent:

220 [decimal] = 11011100 [binary]

In this example, the binary value is a lengthy eight digits versus three digits for the decimal representation. Therefore, a more practical means for dealing with the binary states of digital logic is through a higher-level number system.

While the decimal or base-10 number system is more comprehensible to the human user, the *octal* (base-8) and *hexadecimal* (base-16) number systems mesh more easily with the digital circuit's multiples of four data and address requirements. The handling of these data and address conditions is

Table 1-1. A Comparison Between Four Different Numbering Systems.

Decimal	Binary	Octal	Hexadecimal
0	0	0	0
1	1	1	1
2	10	2	2
3	11	3	3
4	100	4	4
5	101	5	5
6	110	6	6
7	111	7	7
8	1000	10	8
9	1001	11	9
10	1010	12	A
11	1011	13	B
12	1100	14	C
13	1101	15	D
14	1110	16	E
15	1111	17	F
16	10000	20	10

principally executed by several one-digit binary switching circuits, or through a single multi-digit binary register. In an average digital circuit, these registers range in size from 4 to 64 binary digits. This dependence on factors of four makes translating between octal and hexadecimal values and binary digits a necessary talent that must be acquired by the digital circuit designer. Even so, the use of octal programming has recently fallen into disfavor, and the hexadecimal number system has become a veritable standard in the standard-less microcomputer industry.

As an exercise in number system manipulation, perform the following experiment:

Purpose: Write a number system conversion program.

Materials: Any microcomputer system with a high-level language interface (e.g., BASIC, C, and Pascal).

Procedure:
♣ Make your program short with a limited number of program jumps.
♣ Your program should be able to calculate any bit-size binary value.
♣ Test your final program with known binary numbers and compare your achieved results with your acquired results.

Results:

♣ This is one possible solution to this experiment in TurboBASIC code.

```
REM BINARY CALCULATOR PROGRAM
START:
  INPUT "HOW MANY DIGITS ";A
    B=0:C=0:D=0:Z=0:PRINT "ENTER YOUR";A;"-DIGIT BINARY
NUMBER": INPUT QB$
    D=A-1
    FOR B=1 TO A
    IF MID$(QB$,B,1)="1" THEN LET C=Z^D:Z=Z+C:D=D-1
    IF MID$(QB$,B,1)="0" THEN LET C=0:Z=Z+C:D=D-1
    NEXT B
    PRINT QB$;"="
    PRINT TAB(A+1) Z
CHECK:
  INPUT "DO ANOTHER: Y OR N ? ";X$
    IF X$="Y" THEN GOTO START
    IF X$="N" THEN END
    IF X$<>"Y" OR X$<>"N" THEN GOTO CHECK
```

♣ After you have completed this experiment, add an octal and hexadecimal conversion option to your final program.

Referring to a register size as "multi-digit binary" can be almost as cumbersome as writing large binary numbers. Therefore, another means for expressing binary digits is with bits. A bit can equal either a 0 or a 1. Applying this definition to the previously mentioned 8-digit binary number example, 11011110, yields an 8-bit number. When dealing with register bit size, however, the final value can represent the computational strength of a microcomputer.

At the heart (or brain) of every microcomputer is the MPU. This single chip or IC (Integrate Circuit) is the repository of the CPU's registers. Based on the bit-width of these registers, the data handling ability of the MPU can be fairly judged; or, can it? Unfortunately, not all registers are created equal, and many MPUs are difficult to pigeonhole into an accurate statement of their true computing power. Take the Motorola MC68000 MPU, for example.

The MC68000 has 32-bit internal registers, but this same MPU also has a 16-bit data bus, a 23-bit address bus, and a 16-bit *ALU* (Arithmetic Logic Unit). Furthermore, this IC is able to address 16M bytes of unsegmented memory with a 32-bit program counter. So where does this mixed bag of bit-width leave us? Granted, the MC68000 is able to handle 32-bit sized instructions, internally. On the other hand, it can only receive data in 16-bit slices. Therefore, the MC68000 is best classified as a $^{16}/_{32}$-bit MPU.

Another MPU that falls into this dual personality mold is the Intel 8088. The 8088 is best labeled an $^{8}/_{16}$-bit MPU. In this case, there are 16-bit internal registers with an 8-bit data bus, a 20-bit address bus, and 16-bit instruction pointer (the function of the instruction pointer is similar to the role of the program counter in the MC68000). Additionally, the ALU is 16 bits wide. Once again, however, the 16-bit 8088's ability to receive data in 8-bit pieces mandates the $^{8}/_{16}$-bit qualifier in describing its computational strength.

In addition to describing the amount of data that can be processed, the bit-size of a register also indicates the number of logical states that are possible. All of the logic states for two different registers are listed in Table 1-2. Based on the states listed in these two examples, an X-bit register will contain 2^x logic states.

Yet another means for expressing the computational strength of a microcomputer is in its CPU's (Central Processing Unit) memory size. Memory size is usually measured in 8-bit segments. Each 8-bit chunk is defined as a byte. Bytes, in turn, are frequently described in terms of kilobytes (K bytes), megabytes (M bytes), and gigabytes (G bytes). A 1K-byte CPU memory is equal to 2^{10} or 1024 bytes. This same 1K-byte CPU memory also contains 8192 bits (8 × 1024 bytes). Continuing with these equivalencies:

1M bytes = 2^{20} or 1048576 bytes
1G bytes = 2^{30} or 1073741824 bytes

In turn, these same values equal:

1M bytes = 8388608 bits
1G bytes = 8589934592 bits

2-bit Register	
bit 1	bit 0
0	0
0	1
1	0
1	1

3-bit Register		
bit 2	bit 1	bit 0
0	0	0
0	0	1
0	1	0
0	1	1
1	0	0
1	0	1
1	1	0
1	1	1

Table 1-2. The Logic States for Multi-Input Registers.

Bit strings of a byte's width are also useful for indicating the CPU's status. Depending on its register location, a byte can have a variety of different meanings. For example,

00101111

which is equal to,

47 [decimal]
2F [hexadecimal]
57 [octal]

can represent the 8088 instruction code DAS. This code is a mnemonic representation of "decimal adjust for subtraction." One final interpretation of this byte is found in its ASCII form. *ASCII* (American Standard Code for Information Interchange) is a coding system that uses 7 bits of data for describing alphabetic, numeric, and punctuation characters in a digital circuit. Continuing with our previously mentioned byte,

00101111

represents the ASCII character

/

Another alphanumeric character description code is *EBCDIC* (Extended Binary Coded Decimal Interchange Code). This code is used primarily on mainframe IBM systems such as the IBM 360/370. In addition to being restricted to mainframe usage, EBCDIC data is 8 bits wide. Therefore, the byte, 00101111, is equal to the EBCDIC control character BEL or bell.

BINARY MATHEMATICS

In addition to these possible interpretations of an 8-bit number, there are several binary coding schemes that are used in digital circuits. The four most popular codes are: Binary Coded Decimal, Excess 3, Gray, and 2-Out-of-5. Each of these codes use a unique binary format for representing decimal numbers.

Binary Coded Decimal

Binary Coded Decimal (BCD) or 8421 codes are used for translating any decimal number into a binary string. In BCD code, only four binary digits are used. These four bits equal the decimal values 0-9.
For example,

8 [decimal] = 1000 [BCD]

Please note that this binary value is not the same as the 7-bit ASCII code for the decimal number 8. In this case, the 7-bit ASCII code for 8 [decimal] is equal to 0111000.

Even multi-digit decimal numbers can be converted into BCD code. This conversion, however, results in some interesting number system oddities.

For example,

246 [decimal] = 0010 0100 0110 [BCD]

Likewise,

0010 0100 0110 [BCD] = 246 [hexadecimal]

In this example, both the decimal, and hexadecimal numbers for the BCD code are identical. This decimal-hexadecimal equality is a peculiarity found in all BCD code. Based on this equivalence, many mathematicians claim that the BCD code is a subset of the hexadecimal number system.

Excess 3

The Excess 3 code shares many of the same features as BCD code. The only difference between these two binary coding schemes is that Excess 3 code has the value 3 added to each decimal digit prior to BCD code conversion.

For example,

246 [decimal] = 0010 0100 0110 [BCD] = 0101 0111 1001 [Excess 3]

Gray

Binary strings can also be used in error detection. The Gray code is an error detection method where only one binary digit is changed in successive numbers.
For example,

8 [decimal] = 1100 [Gray]

and

246 [decimal] = 0011 0110 0101 [Gray]

2-Out-of-5

Another error detection code is 2-Out-of-5. In this coding scheme, each decimal number is represented by 5 bits. These 5 bits must have two ones and three zeros with a different arrangement for each number. One of the most frequent error detection systems used with 2-Out-of-5 code is checking for even parity. This error detection means is effected by the ever present two ones.

For example,

8 [decimal] = 10100 [2-Out-of-5]

and

246 [decimal] = 00110 01010 10001 [2-Out-of-5]

Once a binary string has been placed in an MPU's register, two basic arithmetic operations can be performed: addition or subtraction. The two states of binary numbers, 0 and 1, make the execution of these functions simpler than the comparable operation in a larger number system.

Binary Addition

In binary addition, there are two possible outcomes: non-carry and carry. A carry in binary addition will only occur when two 1's are added together. This carry will be a 1.

For example,

$$0 + 0 = 0$$
$$0 + 1 = 1$$
$$1 + 0 = 1$$
$$1 + 1 = 0 \text{ with a 1 carry}$$
$$0011 + 0111 = 1010$$

Binary Subtraction

Two forms of number complements are used in binary subtraction: the ones- (1's) complement and the twos- (2's) complement. These complements convert the negative value into a positive format so that binary addition can be used for arriving at the difference. In this manner, both binary addition and binary subtraction can be performed in the same digital circuit.

The Ones-Complement. This is a simple "state-reversing" procedure. In other words, each 1 is changed to a 0, and each 0 is inverted to a 1.

For example,

$$0011 = 1100 \text{ [ones-complement]}$$
$$0111 = 1000 \text{ [ones-complement]}$$

Binary subtraction is performed by taking the ones-complement of the negative value and adding it to the other binary number. Any carry is added to the final sum.

For example,

$$0111 - 0011 = 0111 + 1100 = 1\ 0011 \text{ [1 carry]}* = 0011 + 1 = 0100$$

*—a carry is usually represented as a separate 1 in a column to the left of the most significant digit (furthest left).

The Twos-Complement. Following the ones-complement inversion, a one is added to the least significant bit (furthest right). This addition follows the same carry rules used in binary addition.

For example,

$$0011 = 1101 \text{ [twos-complement]}$$
$$0111 = 1001 \text{ [twos-complement]}$$

Two conventions that are used in twos-complement:

❖ A 0 is placed in front of positive, signed numbers.
❖ A 1 is placed in front of negative, signed numbers.

Once the twos-complement of the negative number has been determined, binary subtraction is performed by adding the twos-complement to the other binary number. In certain cases, a carry will be executed from the most significant digit (the furthest left). This carry is dropped in twos-complement binary subtraction.

For example,

$$0111 - 0011 = 0111 + 1101$$
$$= 1\ 0100\ [1\ carry]$$
$$= 0100$$

BINARY LOGIC

Taking this foundational knowledge of binary mathematics and applying it to digital circuits requires an understanding of binary logic, and its associated digital component equivalents. Basically, binary logic can be broken down into four elements:

+ gates
+ Boolean algebra
+ logic families
+ logic circuits

Gates

There are three general gates in binary logic: the AND gate, the OR gate, and the NOT gate. Each of these gates duplicates a binary mathematical operation. For example, The AND gate represents binary multiplication:

$$INPUT\ A \times INPUT\ B = OUTPUT\ C$$

The OR gate represents binary addition:

$$INPUT\ A + INPUT\ B = OUTPUT\ C$$

The NOT gate represents binary ones-complement:

$$INPUT\ A = -OUTPUT\ C$$

Note: The NOT gate is sometimes referred to as the binary negation operation or the inverter gate, and is formally represented as: $INPUT\ A = OUTPUT\ \overline{C}$ (pronounced "not C").

Special logic symbols are used to represent each of these gates (see Fig. 1-1). These symbols are particularly useful when establishing the Boolean algebra of a given logic circuit.

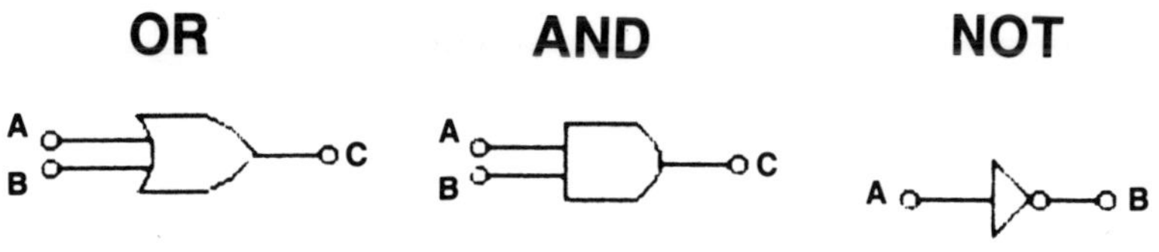

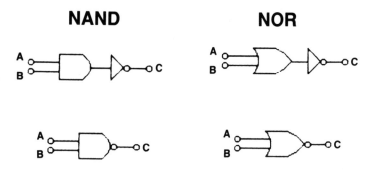

Fig. 1-1. AND, OR, and NOT logic gate symbols.

Fig. 1-2. Creating NAND and NOR logic gate symbols.

The inverter gate can be combined with the output of AND and OR gates to form special functions (see Fig. 1-2). These newly created gates are called the NOT AND or NAND and NOT OR or NOR gates, respectively (the negation operation of the inverter supplies the "NOT" or "N" prefix in the NAND and NOR gates). Likewise, the mathematics of these two new gates are the opposite of their positive logic counterparts.

The NAND gate represents negated binary multiplication:

$$INPUT\ A \times INPUT\ B\ =\ -OUTPUT\ C$$

The NOR gate represents negated binary addition:

$$INPUT\ A + INPUT\ B\ =\ -OUTPUT\ C$$

Just as the inverter gate can be applied to the output of AND and OR gates, this logic reverser can also be combined with either or both of the gate's inputs (see Fig. 1-3).

The AND gate represents negated-input:

$$-INPUT\ A \times INPUT\ B\ =\ OUTPUT\ C$$
$$INPUT\ A \times -INPUT\ B\ =\ OUTPUT\ C$$

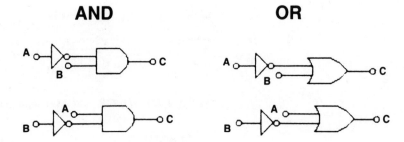

Fig. 1-3. Applying NOT gates to AND and OR inputs.

The OR gate represents negated-input:

$$-\text{INPUT A} + \text{INPUT B} = \text{OUTPUT C}$$
$$\text{INPUT A} + -\text{INPUT B} = \text{OUTPUT C}$$

More than two inputs can be attached to each of these gates and gate combinations. Even with these multiple inputs, the operation of the gate remains unchanged. For example,

The AND gate:

$$\text{INPUT A} \times \text{INPUT B} \times \text{INPUT C} = \text{OUTPUT D}$$

The NOR gate:

$$\text{INPUT A} + \text{INPUT B} + \text{INPUT C} + \text{INPUT D} = -\text{OUTPUT E}$$

One final gate that is used in binary logic is the EXCLUSIVE OR gate. Superficially, the operation of the EXCLUSIVE OR and the OR gate look identical. On further Boolean algebraic scrutiny, however, the final output value will be different for both logic functions.

The EXCLUSIVE OR gate represents exclusive binary addition:

$$\text{INPUT A} + \text{INPUT B} = \text{OUTPUT C}$$

The EXCLUSIVE OR gate can also be negated by an inverter gate. This NOT gate is placed on the output of the EXCLUSIVE OR gate and forms an EXCLUSIVE NOR gate.

The EXCLUSIVE NOR gate represents negated exclusive binary addition:

$$\text{INPUT A} + \text{INPUT B} = -\text{OUTPUT C}$$

Boolean Algebra

The logical combination of the two binary states is the basis for Boolean algebra. Every Boolean algebraic operation is expressed in a truth table. When combined with the logic symbols and mathematical operations formulas, the truth table forms a complete picture of the logic contained in a particular circuit. It is even possible to derive any two of these logic members given the information found in the remaining one. For example:

AND

A + B = C

A	B	C
0	0	0
0	1	0
1	0	0
1	1	1

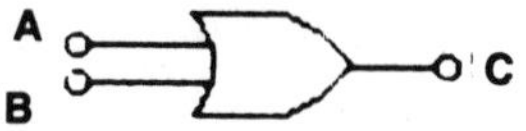

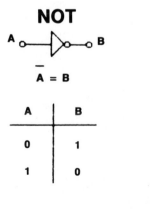

Fig. 1-4. AND logic symbol, logic formula expression, and truth table.

OR

A × B = C

A	B	C
0	0	0
0	1	1
1	0	1
1	1	1

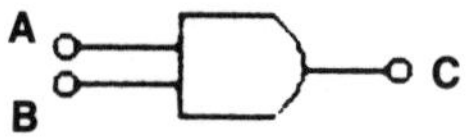

Fig. 1-5. OR logic symbol, logic formula expression, and truth table.

NOT

$\overline{A} = B$

A	B
0	1
1	0

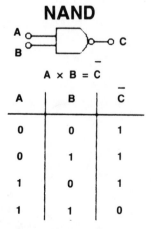

Fig. 1-6. NOT logic symbol, logic formula expression, and truth table.

NAND

$A \times B = \overline{C}$

A	B	$\overline{C}$
0	0	1
0	1	1
1	0	1
1	1	0

Fig. 1-7. NAND logic symbol, logic formula expression, and truth table.

Negated-Input

NOR

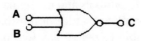

$$A + B = \overline{C}$$

A	B	C
0	0	1
0	1	0
1	0	0
1	1	0

Fig. 1-8. NOR logic symbol, logic formula expression, and truth table.

Negated-Input

AND

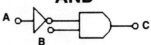

$$\overline{A} \times B = C$$

A	B	C
0	0	0
0	1	1
1	0	0
1	1	0

Fig. 1-9. Negated-input AND logic symbol, logic formula expression, and truth table.

Negated-Input

OR

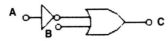

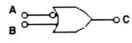

$$\overline{A} = B = C$$

A	B	C
0	0	1
0	1	1
1	0	0
1	1	1

Fig. 1-10. Negated-input OR logic symbol, logic formula expression, and truth table.

Multiple-Input

AND

$$A \times B \times C = D$$

A	B	C	D
0	0	0	0
0	0	1	0
0	1	0	0
0	1	1	0
1	0	0	0
1	1	0	0
1	1	1	1

Fig. 1-11. Multiple-input AND logic symbol, logic formula expression, and truth table.

Multiple-Input

OR

Exclusive

OR

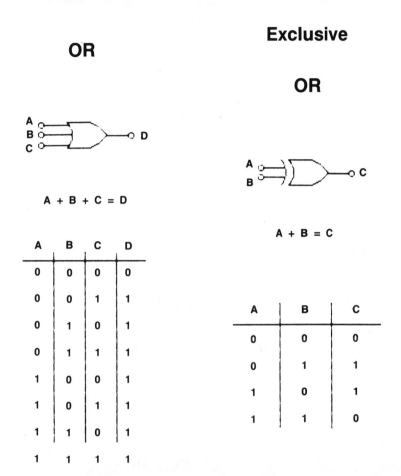

$$A + B + C = D$$

A	B	C	D
0	0	0	0
0	0	1	1
0	1	0	1
0	1	1	1
1	0	0	1
1	0	1	1
1	1	0	1
1	1	1	1

$$A + B = C$$

A	B	C
0	0	0
0	1	1
1	0	1
1	1	0

Fig. 1-12. Multiple-input OR logic symbol, logic formula expression, and truth table.

Fig. 1-13. Exclusive OR logic symbol, logic formula expression, and truth table.

An interesting footnote to the EXCLUSIVE OR truth table is that any binary number that is EXCLUSIVE-ORed with itself will result in all zeros. In other words, a binary number of length x-bits when EXCLUSIVE-ORed with itself will generate an x-bit length binary number filled with zeros. For example,

[EXCLUSIVE OR]   11011011 + 11011011 = 00000000

Once the interrelationship between the truth table, the mathematical operation formula, and the logic symbol is understood, complex multi-gate circuits can be devised. The creation of these multiple-gate circuits can be derived from any one of these binary logic members. For example, observe Figs. 1-14 through 1-17.

Based on these examples, two generalized methods of deriving the mathematical operations formula from the truth table can be easily described. In the first method, form a product on each row that has a one in the output column. A negation must be used for each zero input column. Next, add each of these products together. The result is the mathematical operations formula for the particular truth table. This derivation method is sometimes called the sum-of-products rule. See Fig. 1-18.

The second method for learning the mathematical operations formula based on a truth table is called the product-of-sums rule. Basically, this method is just the reverse of the sum-of-products rule. In product-of-sums, form a sum on each row that has a zero in the output column. Be sure to use a negation for each one input column. Finally, multiply all of these sums together. See Fig. 1-19.

As a rule, use whichever derivation method that will provide the simplest results. This determination can be based on the number of ones in the output

OR Gate

AND Gate

AND Function

A	B	C
0	0	0
0	1	0
1	0	0
1	1	1

Fig. 1-14. Multi-gate OR with an AND logic gate function.

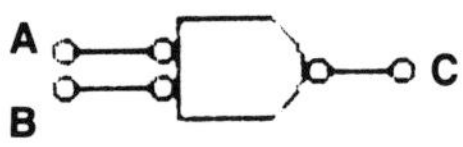

OR Function

A	B	C
0	0	0
0	1	1
1	0	1
1	1	1

Fig. 1-15. Multi-gate AND with an OR logic gate function.

Four Inputs, One Output

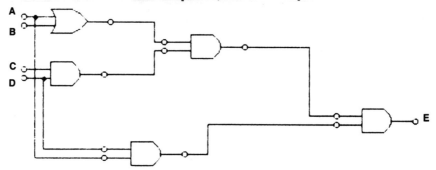

Fig. 1-16. Four inputs with one output.

Three Inputs, Two Outputs

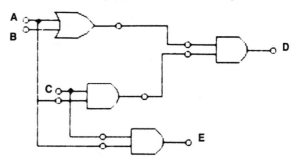

Fig. 1-17. Three input with two outputs.

Sum-of-Products

$$\overline{A} \times \overline{B} + \overline{A} \times B + A \times \overline{B} = \overline{A}\overline{B} + \overline{A}B + A\overline{B} = C$$

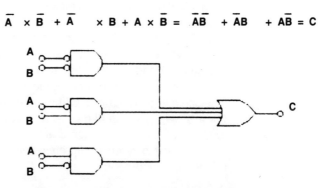

Fig. 1-18. An example of the sum-of-products rule.

Product-of-Sums

$$(\overline{A} + \overline{B}) \times (\overline{A} + B) \times (A + \overline{B}) = C$$

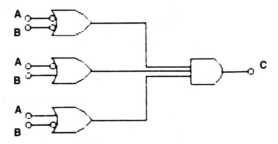

Fig. 1-19. An example of the product-of-sums rule.

column. In other words, if more ones than zeros are in the output column, then use product-of-sums. On the other hand, if more zeros are present in the output column, then use the sum-of-products rule.

More often than not, the mathematical operation formulas can get too large and awkward for easy translation into logic symbols. To help in minimizing this congestion, there are several Boolean rules and laws that can be used for reducing, manipulating, and simplifying these binary formulas (see Fig. 1-20).

Applying Boolean algebra to actual digital circuits requires only one minor alteration. Instead of dealing with 0's and 1's, digital circuits deal with voltages. These voltages usually come in two forms: 0 volts and +5 volts. Coincidentally, these two voltage levels are ideal representatives of the two binary states. By setting 0 volts equal to a binary 0 and +5 volts equal to a binary 1, a typical positive logic circuit can be formed.

Conversely, a negative logic circuit is possible by setting 0 volts equal to a binary 1 and +5 volts equal to a binary 0. This positive-to-negative logic conversion practice also changes the function of the basic gates. In other words, a negative logic AND gate will function like an OR gate and a negative logic OR gate will function like an AND gate. Based on this information, only the function of the gate is altered and not its circuitry.

As a test of your comprehension of binary mathematics, perform the following experiment:

Purpose: Simplify a complex binary mathematical operation formula.

Boolean Rules and Laws

Rule 1	$A \times 0 = 0$
Rule 2	$A \times 1 = A$
Rule 3	$A \times A = A$
Rule 4	$A \times \overline{A} = A$
Rule 5	$\overline{\overline{A}} = A$ (Tandem NOTs)
Rule 6	$A + 0 = A$
Rule 7	$A + 1 = 1$
Rule 8	$A + A = A$
Rule 9	$A + \overline{A} = 1$
Rule 10	$A + AB = A$
Rule 11	$A \times (A + B) = A$
Rule 12	$(A + B)(A + C) = A + BC$
Rule 13	$A + \overline{A}B = A + B$
Cummutative Law 1	$A + B = B + A$
Cummutative Law 2	$AB = BA$
Associative Law 1	$A + (B + C) = (A + B) + C$
Associative Law 2	$A(BC) = (AB)C$
Distributive Law	$A(B + C) = AB + AC$
Demorgan's Law	$\overline{A + B} = \overline{A}\overline{B}$ or $\overline{AB} = \overline{A} + \overline{B}$

Fig. 1-20. The laws and rule for use with binary mathematics.

Procedure:
- ❖ Use the rules and laws of Boolean algebra to derive and simplify the binary mathematical operation formula based on the given truth table.
- ❖ Draw the logic symbols for the solved formula.
- ❖ Write a high-level language program for solving complex binary mathematical formulas.

A	B	C
0	0	1
0	1	1
1	0	0
1	1	0

1. What is the simplest formula for this truth table?

2. What is the logic diagram for this truth table?

$$\overline{A} + \overline{B} + (\overline{AB}) + A + \overline{AB} = C$$

1. What is the truth table for this equation?

2. What is the logic diagram for this equation?

Fig. 1-21. Answer these questions based on the supplied logic information.

Logic Families

Each of the binary logic gates have been duplicated in special digital ICs. These chips represent varying degrees of processing speed, power consumption, and construction material. Based on these different qualities, several large digital IC groupings can be made. These IC groupings are called logic families. In general, there are 15 logic families: primitive switch logic, switching mode logic, resistor-transistor logic, diode-transistor logic, fan-out logic, transistor-transistor logic, Schottky TTL, emitter-coupled logic, integration-injection logic, MOS logic, NMOS logic, PMOS logic, VMOS logic, CMOS logic, and QMOS logic.

Primitive Switch Logic. Early attempts at logic gates were made with analog switches. Both AND and OR gates were easily duplicated with these SPST (Single-Pole, Single-Throw) switches (see Fig. 1-22). A slight increase in the gate's switching function was made possible by substituting diodes for these switches (see Fig. 1-23). Even NPN transistors (a bipolar N-type, P-type, N-type transistor) could be added to this diode gate for generating a NOT gate (see Fig. 1-24).

Switching Mode Logic. Eventually, all gates were constructed from bipolar transistors. There are three types of bipolar switching mode logic gates: current sourcing, current sinking, and current mode logic.

In *current sourcing,* all subsequent gate transistor base current is derived from the previous gate's collector (see Fig. 1-25). This switching mode logic type is frequently used in Resistor-Transistor logic.

AND

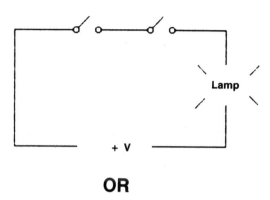

OR

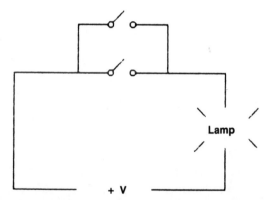

Fig. 1-22. AND and OR gates constructed from simple SPST switches.

Current sinking, on the other hand, draws the second gate's base current from a bias resistor that is attached between its own base and collector (see Fig. 1-26). The sink for the second gate, however, is produced by the collector of the previous gate.

Finally, *current mode logic* or CML uses a current limiting resistor to control the activation of a given bipolar transistor depending on the state of the input signal (see Fig. 1-27). CML gates operate at a higher frequency than any of the previously mentioned logic families. Even some of the later logic circuits operate at a lower frequency due to the saturation requirement of their transistors.

AND

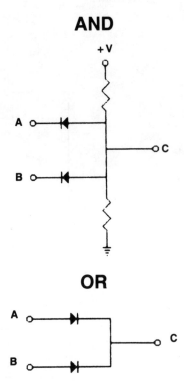

OR

Fig. 1-23. AND and OR gates constructed from diodes.

NOT

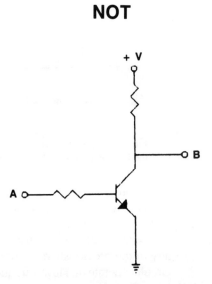

Fig. 1-24. A NOT gate constructed from two resistors and an NPN transistor.

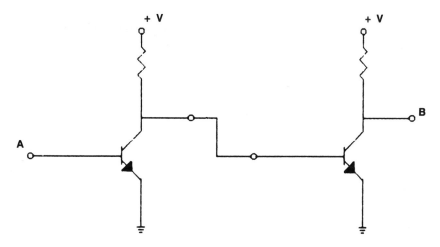

Fig. 1-25. A current sourcing circuit.

Resistor-Transistor Logic. Resistor-transistor logic or RTL differs from CML by placing a resistor on the gate's transistor collector, as well as a current limiting resistor on the transistor's base (see Fig. 1-28). By adding a capacitor to the current limiting base resistor, a resistor-capacitor-transistor logic or RCTL gate can be created (see Fig. 1-29). The advantage of the RCTL gate is an increase in the maximum operating frequency.

Diode-Transistor Logic. Controlling the voltage drop of an RTL gate through input diodes is diode-transistor logic or DTL (see Fig. 1-30). DTL is accomplished by fixing diodes to the transistor's base.

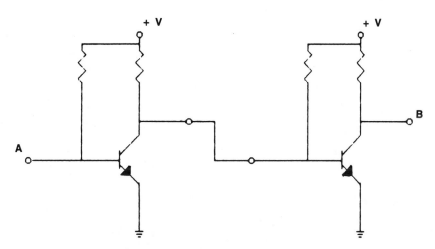

Fig. 1-26. A current sinking circuit.

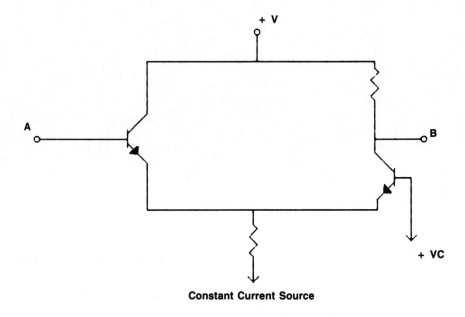

Constant Current Source

Fig. 1-27. A current mode logic circuit.

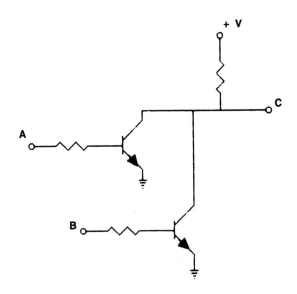

Fig. 1-28. An RTL circuit.

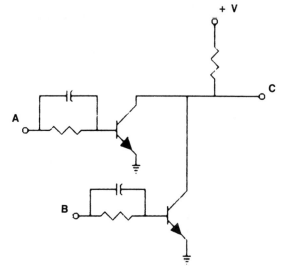

Fig. 1-29. An RCTL circuit.

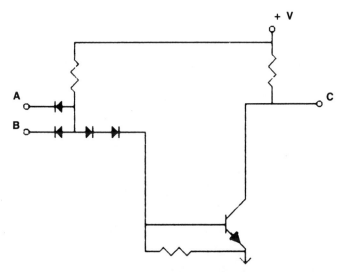

Fig. 1-30. A DTL circuit.

Fan-Out Logic. In a digital circuit, the binary states of 0 and 1 are depicted as voltage changes. Ideally, these states are 0 volts for a zero bit and +5 volts for a one bit. More typically, however, these voltage variations are 0 to +0.7 volts and +2.0 to +5.0 volts for a binary 0 and 1, respectively. By representing these binary states with two such sweeping voltage ranges, errors in interpretation can occur. These problems arise when adding additional digital circuitry to a gate's output. In actual operation, each added circuit lowers the output current of the gate. Once the output of the gate reaches a level lower than +2.0 volts, the logic of the gate will fail to function correctly. Usually, it takes several added digital circuits to lower a gate's output current sufficiently to destroy its logic. The number of digital circuits that is necessary to lower a gate's output current is the gate's *fan-out value*. For example, if three circuits are attached to a gate's output and the current falls below +2.0 volts, then this gate has a fan-out of three. Understanding a gate's fan-out number will be an important design specification in several of the following logic families.

Transistor-Transistor Logic. Transistor-transistor logic or TTL is one of the largest logic families (e.g., the 74XXX series of digital ICs). TTL gates use current-sinking switching logic mode in their construction (see Fig. 1-31). Another key feature in a TTL gate is the use of transistors with multiple emitters. This gives the gate high speed, low power consumption, and a degree of signal noise immunity.

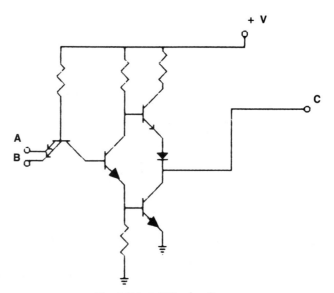

Fig. 1-31. A TTL circuit.

Schottky TTL. By adding a Schottky diode between the base and collector of every TTL transistor, an increase in operating speed and a decrease in power consumption is possible (see Fig. 1-32). Schottky TTL gates operate at a higher frequency by using the Schottky diode as a conductor of bias current, and therefore, reduce the transistor's need for saturation.

Emitter-Coupled Logic. By using current switching and emitter amplifiers, the emitter-coupled logic or ECL gate is able to avoid the transistor saturation

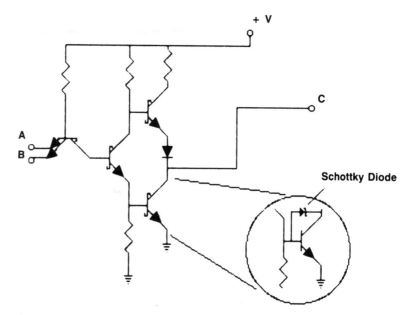

Fig. 1-32. A TTL circuit enhanced with Schottky diodes.

requirement (see Fig. 1-33). An unfortunate side effect of ECL gates is their constant drain on the circuit's power supply.

Integrated-Injection Logic. Integrated-injection logic or IIL transistors employ common bases, emitters, and collectors (see Fig. 1-34). Based on this combining of transistor elements, IIL is also known as MTL or merged-transistor logic. IIL gates have both a high switching speed and a high construction density. There are two construction methods, however, that enhance the performance characteristics of the IIL gate.

A Schottky diode can be added to IIL transistors for an increase in operating speed. This Schottky IIL gate offers many of the same advantages found in the Schottky TTL gate.

The other construction enhancement method is achieved through the use of an *isoplanar ion implantation process*. This isoplanar-integrated-injection logic or IIIL gate has an oxide barrier for isolating its transistors from disruptive electron contamination.

MOS Logic. Other than bipolar transistors, metal-oxide-semiconductor field-effect transistors or MOSFETs can be used in gate construction. Basically, a MOSFET has a metal gate that is insulated from a silicon semiconductor substrate by a silicon dioxide shield (see Fig. 1-35). This is MOS (metal-oxide-semiconductor) technology. The current flows between the source and the drain. This flow is controlled by a voltage that is applied to the metal gate. There are four typical arrangements of these components within a MOSFET: N-channel, depletion mode, P-channel, depletion mode, N-channel,

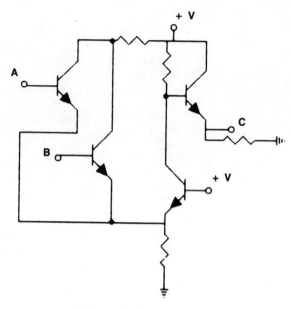

Fig. 1-33. An ECL circuit.

enhancement mode, and P-channel, enhancement mode. The major difference between a depletion mode MOSFET and an enhancement mode MOSFET is the presence of a conducting channel running between the source and the drain in the depletion mode. This channel is etched in the semiconductor substrate and it is able to conduct a 0 volt gate state. Enhancement mode MOSFETs lack this channel.

NMOS Logic. N-channel MOSFET or NMOS gates have an N-type source and drain in a P-type substrate (see Fig. 1-36). Applying voltage to the metal gate forms a current between the source and drain.

PMOS Logic. Converse to an NMOS gate, P-channel MOSFET or PMOS gates have a P-type source and drain in an N-type substrate (see Fig. 1-37).

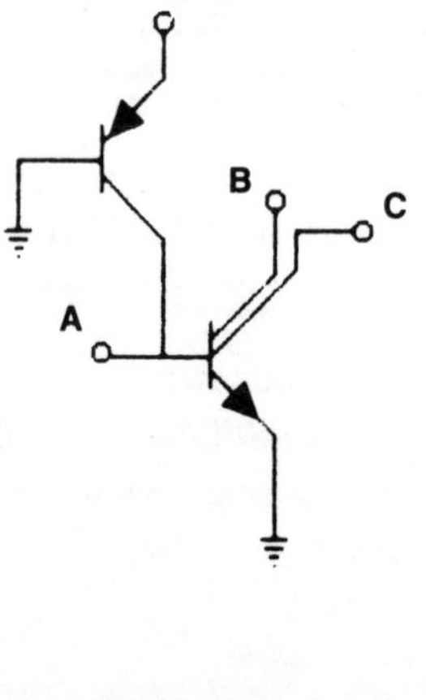

Fig. 1-34. An IIL circuit.

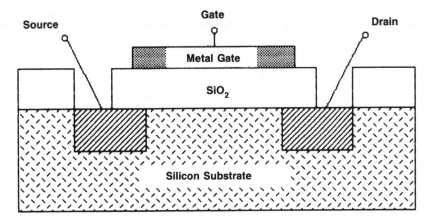

Fig. 1-35. A cross-section of a MOSFET circuit.

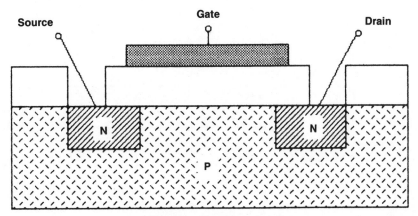

Fig. 1-36. An NMOS circuit.

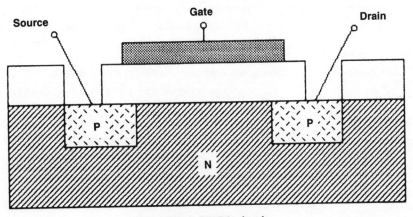

Fig. 1-37. A PMOS circuit.

A PMOS gate operates just like an NMOS one, except that the PMOS has an opposite polarity.

VMOS Logic. VMOS or vertical MOSFET gates use a unique construction technique that provides the "vertical" portion of its name (see Fig. 1-38). VMOS gates have an increased power capability over NMOS and PMOS gates. This technology is frequently used in high-density RAM (random access memory) chips.

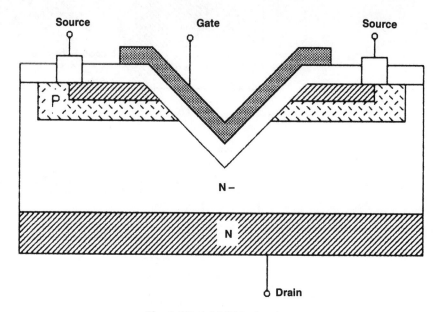

Fig. 1-38. A VMOS circuit.

CMOS Logic. Combining both N-channel and P-channel MOSFETs into the same gate is complementary MOS or CMOS logic (see Fig. 1-39). CMOS gates have a very low power consumption with a wide range in voltage tolerance. Accompanying this reduced power consumption, however, is the lack of speed. For example, a typical CMOS gate is slower than a similar TTL gate. Furthermore, a CMOS gate has a high input impedance. A drawback to this heightened input impedance is the CMOS gate's susceptibility to static electrical discharge.

QMOS Logic. High-speed CMOS or QMOS logic is both functionally identical and pin compatible with 74XXX TTL ICs (see Fig. 1-40). QMOS gates require a slightly elevated current, but the consumption is less than that found in Schottky TTL gates. On the other hand, the operating speed of the QMOS gate is approximately 10 times faster than that of the comparable CMOS gate. In fact, the QMOS gate's speed is close to that of the Schottky TTL gate with less power drain.

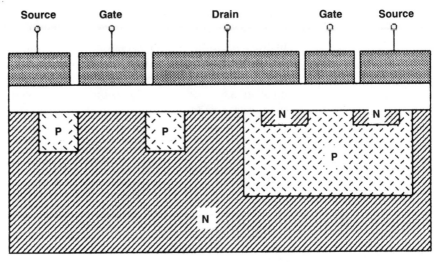

Fig. 1-39. A CMOS circuit.

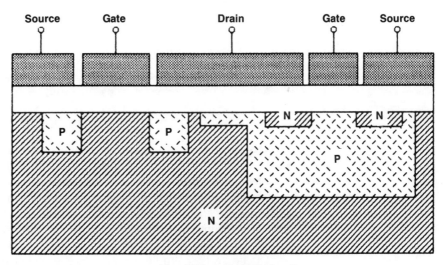

Fig. 1-40. A QMOS circuit.

Logic Circuits

On a practical level, a logic circuit is a direct application of all of the previously discussed gates and logic families. From this manufacturing of AND, OR, and NOT gates, with a particular logic family fabrication technique, a logic IC is created. Generally speaking, there are 12 major categories of logic circuits: arithmetic, decoders, latches, flip-flops, multivibrators, counters, shift registers, multiplexers, demultiplexers, drivers, bilateral switches, and tri-state logics.

Arithmetic. Arithmetic logic circuits perform binary addition and subtraction. There are four types of arithmetic circuits for executing this mathematics: half-adder, full-adder, half-subtractor, and full-subtractor. The addition and subtraction of two bits is carried out by a half-adder and a half-subtractor, respectively. By combining two half-adders or two half-subtractors together, a full-adder and a full-subtractor is created. A 4-bit Binary Full Adders with Fast Carry 7483 is an example of an arithmetic logic circuit (see Fig. 1-41).

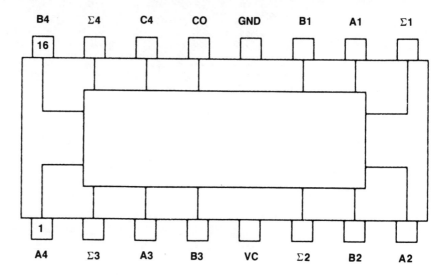

Fig. 1-41. Internal wiring logic for a 7483.

Decoders. Converting between number systems, coded and uncoded variables, and number representations is handled by decoder circuits. A decoder is able to take a given input and change it into a desired output. Some of the more common decoders can change a BCD input into a decimal output, a binary input and change it into an octal output, or use a BCD input to drive a seven-segment LED (Light-Emitting Diode) display. The Excess-3-to Decimal 4-Line-to-10-Line Decoder 7443 is a digital TTL decoder (see Fig. 1-42).

Latches. A latch is a simple memory gate with a feedback loop that toggles the gate's logic between a 0 and 1 state. An Octal D-type Latch 74373 is an IC with eight separate D-type latches (see Fig. 1-43).

Flip-Flops. Combining the gates found in a latch with a circuit set and clear function is the simplest form of a flip-flop. There are four types of flip-flops: S-C, clocked, J-K, and D-type. Each subsequent flip-flop type is an increase in the sophistication over the previous one. For example, by adding a clock input to a S-C (or Set-Clear) flip-flop, a clocked flip-flop is formed. Furthermore, the J-K (or Master-Slave) flip-flop can be enhanced to a D-type flip-flop by adding an inverter between the J and K inputs. The Dual J-K Flip-Flop with Clear 7473 is an example of this logic circuit (see Fig. 1-44).

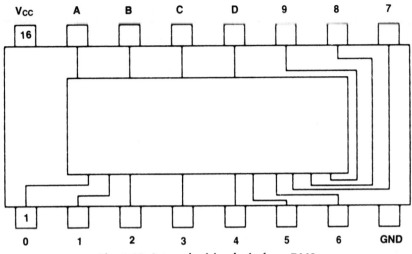

Fig. 1-42. Internal wiring logic for a 7443.

Multivibrators. The circuit's current determines the oscillating frequency of the multivibrator. This determination makes the multivibrator an excellent digital timing device. There are three types of multivibrators: monostable, bistable, and astable. A *monostable multivibrator* has a single stable output state. Similarly, the *bistable* variety has two stable outputs. Both of these multivibrators can be contrasted against the *astable* type which has no stable output states. One problem that can plague multivibrators is fluctuating or ''dirty'' input signals. A *Schmitt Trigger* is a logic circuit that is capable of

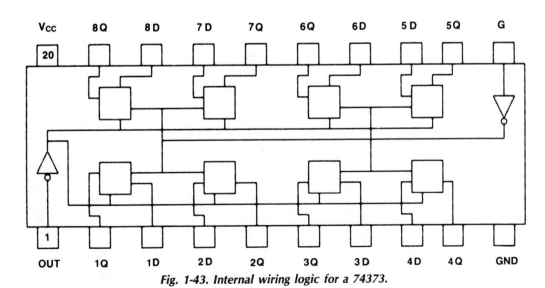

Fig. 1-43. Internal wiring logic for a 74373.

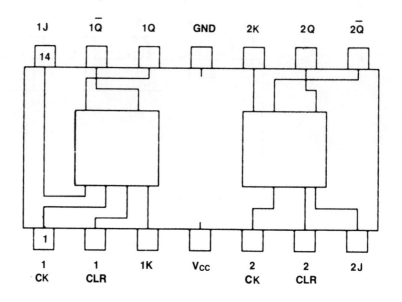

Fig. 1-44. Internal wiring logic for a 7473.

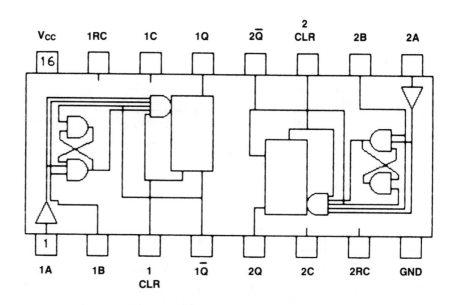

Fig. 1-45. Internal wiring logic for a 74221.

stabilizing these inputs into solid 0 and a 1 logic states. The Dual Monostable Multivibrator 74221 is a TTL example of this circuit (see Fig. 1-45).

Counters. Keeping track of the number of pulses traveling through a circuit is the job of the counter. A simple counter can be made from a series of flip-flop circuits. Dedicated counters are available, however. There are binary, BCD, divide-by-2, and divide-by-12 counter logic circuits. The Divide-by-2 and Divide-by-8 4-bit Binary Counter 7493 is a counter IC (see Fig. 1-46).

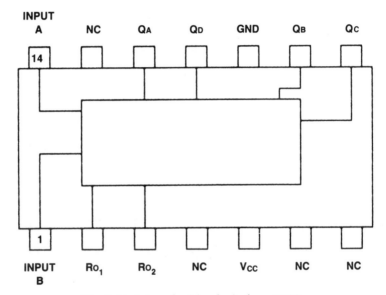

Fig. 1-46. Internal wiring logic for a 7493.

Shift Registers. A hybrid of the digital counter circuit is the shift register. In a standard shift register, a binary number (e.g., 00100101) is moved one place to the left with each clock pulse. During this move zeros are added to the right places as the openings are formed. Therefore, this binary number becomes 01001010 on the first clock pulse and 10010100 on the second pulse. Other digits movements are possible with different shift registers. A 4-bit Parallel-Access Shift Register 74195 is an example of this logic circuit (see Fig. 1-47).

Multiplexers. In a multi-signal circuit, the selective application of several control inputs can determine the nature of the final output signal. This is the purpose of the multiplexer (or MUX) logic circuit. *Multiplexers* are described by the number of total signals that can be selected by its control inputs. In other words, a 4-of-16 multiplexer will select 4 signals from a total of 16 signals based on a given control input. The 3-to-8 Line Decoder/Multiplexer 74138 is a typical multiplexer example (see Fig. 1-48).

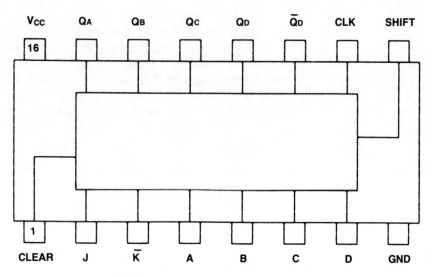

Fig. 1-47. Internal wiring logic for a 74195.

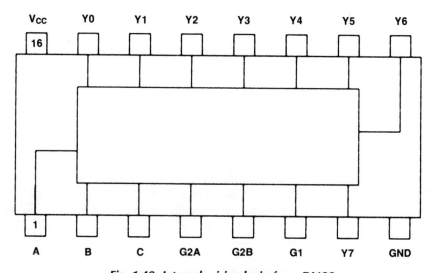

Fig. 1-48. Internal wiring logic for a 74138.

Demultiplexers. A demultiplexer is the opposite of a multiplexer. While a multiplexer established the number of signals that would be applied to a single output, a demultiplexer controls the number of outputs that will be applied to the single input. Similar to the multiplexer, the demultiplexer is named according to the number of outputs that are combined into the input. A Dual 2-to-4 Line Decoder/Demultiplexer 74139 is an example of this logic circuit (see Fig. 1-49).

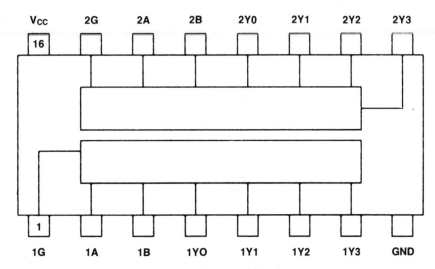

Fig. 1-49. Internal wiring logic for a 74139.

Drivers. Operating a digital display involves the conversion of the circuit's binary states into meaningful display data. This conversion usually involves the changing of 4-bit BCD nibbles into seven-segment LED display data. Frequently these display drivers are combined with BCD decoders in a single IC package. This combination makes a complete conversion and display unit inside a small space. The BCD-to-Seven-Segment Decoder/Driver 7447 is a standard digital LED driver (see Fig. 1-50).

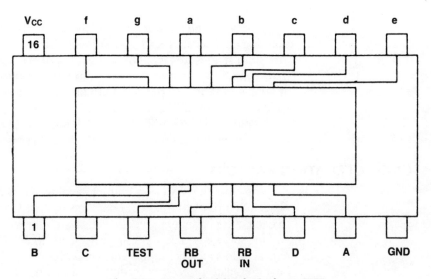

Fig. 1-50. Internal wiring logic for a 7447.

Bilateral Switches. Like the primitive analog switches that were used in early logic gates, there are digital versions available in compact ICs. Contrary to the analog variety, digital switches are controlled through the logic of the circuit. These digital switches lack a fixed polarity and are, therefore, called *bilateral switches*. In other words, each switch lacks a true input and output. A Quad Bilateral Switch for Transmission or Multiplexing of Analog or Digital Signals 4016 is a CMOS example of this logic circuit (see Fig. 1-51).

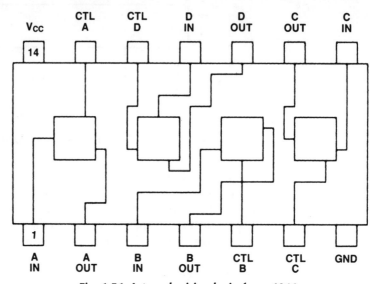

Fig. 1-51. Internal wiring logic for a 4016.

Tri-State Logics. Switching strictly digital signals is performed with tri-state logics. Tri-state logics add another output state to the binary 0 and 1. This third state is no signal or a high impedance absent signal. This special third condition is controlled by an input line. While this other input line is at a logic of 0, the output states are like that of a buffer (i.e., an input of 0, yields an output of 0, etc.). When this control input is a logic 1, however, the output is neither 0 nor 1; it is a high impedance non-signal. The Hex Bus Driver, Noninverted Data Output 74367 is a TTL tri-state logic (see Fig. 1-52).

CMOS OPERATION FACTORS

Implementing CMOS circuitry in analog and digital designs requires the observation of eight different operation factors. Only by faithfully adhering to each of the stringent guidelines that are inherent to each of these factors will the success of a design using CMOS technology be ensured. These eight critical factors include: supply voltage, power dissipation, power source, signal handling, gate protection, system noise, IC handling, and IC interfacing.

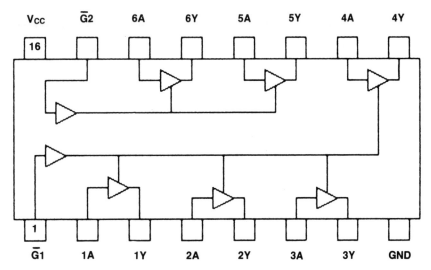

Fig. 1-52. Internal wiring logic for a 74367.

Supply Voltage

As a safeguard against circuit damage, a recommended supply voltage range is provided for all CMOS ICs. Typically, this voltage range is between 3 to 18 volts. The upper limit of this voltage range lies significantly below the minimal voltage limit necessary for achieving primary circuit breakdown. This voltage "buffer" is specified as a precaution against spurious power supply transients which, when added to the supply voltage, would exceed the breakdown limits of the IC.

Power Dissipation

Power dissipation is calculated from the sum of two components: the quiescent component and the dynamic component. The quiescent component is determined from the product of the supply voltage and the surface leakage current summed with the net integrated-circuit reverse diode-junction current. This calculation leads to the dc dissipation of the CMOS device. A standard CMOS IC dc dissipation ranges between 100 to 400 nW for a reference supply voltage of 10 volts. Based on this dc dissipation it is possible to calculate the maximum dc dissipation allowed by a device from the product of the maximum limit quiescent current and the dc supply voltage.

An equation for determining the power dissipation for a CMOS device output buffer is:

$$CV^2f$$

where,

C = load capacitance
V = supply voltage
f = output switching frequency

A derivative of power dissipation is the establishment of the temperature-dependent power rating for each CMOS device. A standard rating of 200 mW is found in CMOS ICs operating at a maximum ambient temperature rating of 85°C to 125°C. Power ratings for temperatures below the maximum rating are graphically portrayed with a thermal derating chart (see Fig. 1-53).

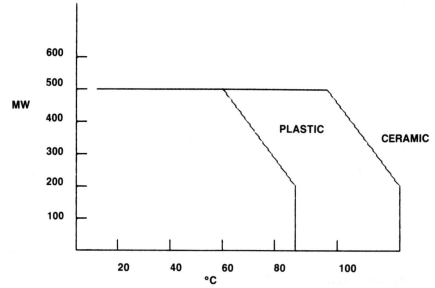

Fig. 1-53. A stylized thermal derating chart for plastic- versus ceramic-cased CMOS devices.

Three design methods for obtaining the optimal degree of thermal derating are:

❖ Mount all CMOS devices in PC sockets or solder each device directly to a graded PCB (Printed Circuit Board).

Maximum Soldering Temperature: + 265°C

❖ Provide unrestricted convection cooling of the installed CMOS devices.
❖ Avoid above standard pressure (>14.7 psi) on the mounted device.

Power Source

There are four different rules that apply to CMOS device power source applications.

❖ Always apply power to the CMOS device prior to the connection of the input signals. This rule is especially true when independent power supplies are used for both the CMOS device and the signal inputs. In this case, the device power supply must be applied first and followed by the signal inputs. Conversely, the signal inputs must be switched off prior to the cessation of the CMOS device's power. Carefully monitoring this power supply priority will prevent damage to the input-protection diode and overdissipation of the device.

❖ Maintain an adequate level below the voltage and current operation limits for the device. This precaution is most true with the power supply voltage and the power source current. In the first case, operate the device well below the absolute maximum supply voltage rating. Likewise, when dealing with the power source current, use the lowest value that is possible without corrupting the operation of the device.

The interfacing of TTL devices with a CMOS device requires an amendment to this rule. In this application, the power supply for the CMOS device must be larger than +5.0 volts. Otherwise, the CMOS device input will exceed V_{DD}.

❖ Observing the correct power supply polarity of the device is vital for protection against a forward-bias current. This current will short the protection diode between V_{DD} and V_{SS}. The maximum limit for polarity protection is a difference of −0.5 volts between the positive (V_{DD}) and negative (V_{SS}) terminals.

Contrasting to this power supply polarity requirement is the relationship between V_{DD} and V_{CC} in dual power supply CMOS buffers. In this case, V_{DD} must be equal to or greater than V_{CC}.

❖ Increased resistance between V_{DD} and V_{SS} will produce transient power applications in the input-protection diodes. Therefore, avoid the installation of multiple resistors in series between these two pins.

Signal Handling

CMOS device signals can be broadly broken down into three major types: input, output, and clock. Along with each of these three signal types, there are separate listings of operation and application rules.

Input Signal Rules:

❖ *All* unused CMOS device inputs must be connected to either V_{DD} or V_{SS} (the destination choice for this connection is dictated by the logic of the device). Failure to observe this rule will result in inaccurate logic operation, exceeding the device's maximum power dissipation rating which will damage the IC.

❖ The voltage of all input signals must be within the specified power supply voltage range. Likewise, the input current limits must be maintained within the tolerance range specified for each device. A typical maximum value for this input current is ± 10 mA. Once again, exceeding this limit can cause the destruction of the device. In this case, the input protection diode will be damaged thereby inducing a $V_{DD} + V_{SS}$ latch condition.

❖ Interfacing TTL and CMOS devices together mandates the implementation of a pull-up resistor between the selected CMOS input and $+5$ volts.

Output Signal Rules:

❖ Avoid all potential output buffer shorts in excess of $+5$ volts. Such a short can occur from large capacitive loads in excess of 5 mF applied directly to buffer outputs.

❖ Carefully monitor the power dissipation of the CMOS device with reference to the specified ambient temperature. This dissipation value can be determined by one of three different procedures:

    1. Short the device outputs directly to either V_{DD} or V_{SS}.
    2. Configure the outputs to drive a low-impedance load.
    3. Configure the outputs to drive the base of a PNP or NPN bi-polar transistor.

❖ Never hard-wire a CMOS device into a "wire-OR" logic condition. This action could short a PMOS or NMOS transistor to the power supply.

❖ Outputs, as well as inputs, may be paralleled only when all of the involved gates are within the same CMOS device.

❖ Following the peak voltage on an output, the output voltage level must fall back inside the specified power supply voltage tolerance range.

❖ Avoid configuring buffer outputs as linear amplifiers, one-shot multivibrators, and astable multivibrators. These configurations will over dissipate the output transistor.

Clock Signal Rules:

❖ Each CMOS device has a maximum clock rise and fall time (5 - 15 microseconds).

✤ Increased rise and fall times will cause a CMOS IC to function incorrectly due to false triggering along with a data ripple effect.

✤ Lengthening the rise and fall times on CMOS buffer inputs will raise the device's power dissipation.

✤ Sequentially clocked ICs should be cascaded with reference to the maximum clock rise time. This value can be determined through the following equation:

$$Ct = (.8 * V_{DD} / k) * t$$

where,

Ct = maximum clock rise time
V_{DD} = power supply in volts
k = a constant 1.25 volts
t = time value (High-to-Low or Low-to-High) in nS

Gate Protection

Based on its inherently high impedance, both the gate oxides and the PN junctions of CMOS devices are extremely sensitive to electrostatic damage. Therefore, several protection diode networks have been engineered into all CMOS devices. The basic gate protection scheme is the resistor-diode network. By increasing the number of diodes and resistors found in this basic network, further improvements and enhancements can be made to the static discharge protection of the device.

System Noise

Noise immunity in a CMOS device is an important application derivative of its balanced low impedance output circuitry. There are, however, three potential sources of system noise that can greatly retard the noise immunity performance of the CMOS IC:

✤ Power supply line noise can be limited through the installation of decoupling capacitors between the power and ground buses. Furthermore, by using separate power supply lines for logic devices versus those used for power switching devices, the negative effects of power supply line noise can be minimized.

✤ AC line noise must be isolated from the CMOS device through transformers and optocouplers.

✤ Ground noise, generated from a shared ground bus, is reduced through the employment of separate ground lines for CMOS logic devices and power switching components.

The presence of the CMOS device's high dc noise immunity is based on the I/O (Input/Output) transfer characteristics. This results in an average 50 percent switching between high and low logic states.

By definition, noise immunity, or noise immunity voltage, is the noise voltage that is isolated on a single input line. The determination of this noise immunity voltage is based on a percentage of the power supply voltage. In other words, the minimum noise immunity voltage for unbuffered CMOS packages is 30 percent to 27 percent with power supply voltage ratings of +5.0 to +10.0 volts and +15.0 volts, respectively. Conversely, buffered CMOS ICs have a minimum noise immunity voltage of 20 percent (regardless of the power supply voltage).

Following the determination of the noise immunity voltage, the final system noise factor, noise margin, can be examined. The *noise margin* (or, noise margin voltage) of a CMOS device is the difference between the noise immunity voltage and the output voltage. This noise margin voltage value represents the maximum voltage that can be applied to any given logical input voltage without disrupting the device's logic or exceeding the output voltage. Under general operating conditions, the minimum noise margin for a buffered CMOS IC ranges between +1.0 to +2.5 volts for a power supply range of +5.0 to +15.0 volts, respectively.

In addition to dc noise immunity, CMOS devices also exhibit ac noise immunity. This value is established through the relationship between the input pulse width and the device's propagation delay. There are two cases where ac noise immunity is important:

❖ Positive noise pulses on the ground bus or on a low signal line.
❖ Negative noise pulses on the power supply bus or on a high signal line.

IC Handling

There are six CMOS device handling areas that require special design precautions:

Humidity. An environment with a relative humidity under 30 percent increases the possibility for production of device damaging electrostatic charges. Therefore, handle CMOS packages in work areas with a minimal relative humidity of 40 percent.

Conductivity. *All* unmounted CMOS ICs should be placed in conductive carriers. This precaution also applies to table tops and work areas.

Grounding. When attaching a CMOS device to a PCB substrate only a grounded soldering iron should be used for making the bond. Likewise, the engineer performing the chip soldering should be grounded through wrist straps and a series resistor (typically 1MΩ).

Shorting. During the transportation of a CMOS package, all external leads should be shorted together.

Soldering. A unique soldering sequence should be used when bonding a CMOS IC to a PCB substrate. Always make the V_{DD}-to-power supply connection, first. Follow this initial connection with the bonding of V_{SS} to ground along with all subsequent lead attachments.

Testing. All CMOS device testing must be performed by a grounded engineer. Use this six-step sequence for testing a CMOS package:

❖ Insert the target IC into its test socket.
❖ Apply the power supply voltage to V_{DD}.
❖ Instigate the input signal.
❖ Test the target device.
❖ Remove the input signal.
❖ Remove the power supply voltage to V_{DD}.

IC Interfacing

Based on its broad range of operating power supply voltages, low power consumption, and low input current, a CMOS device can be interfaced to a large number of different logic families and support components. For example, with the proper conversion circuitry, the following interfaces are possible:

> CMOS-to-DTL
> CMOS-to-ECL
> CMOS-to-TTL
> CMOS-to-NMOS
> CMOS-to-PMOS
> CMOS-to-LED
> CMOS-to-LCD
> CMOS-to-Op Amp
> CMOS-to-Power Loads

Virtually all standard CMOS packages are supplied with directly solderable lead-tin plated chip leads. The use of these leads makes the direct soldering of CMOS ICs to PCB substrates extremely reliable. Once these ICs are connected to the substrate, however, they still retain their special handling characteristics until the subtrate is properly connected to a rated power supply system. During this interface procedure, there are three factors that must be strictly monitored:

1. Electrical performance.
2. Thermal performance.
3. Mechanical performance.

CMOS LOGIC TESTER

Unfortunately, today's digital IC packages hide their logic inside a sterile black plastic housing. Documenting the gates and inverters that are masked into each of these components is usually obtained through the study of the manufacturer's logic diagrams. A superior method for labeling the logic of a digital IC's pins is through the application of the CMOS Logic Tester. This simple project also provides a benefit that can't be found in a manufacturer's documentation—verification of a chips condition. In other words, the CMOS Logic Tester not only indicates the logic of each pin on a digital IC, it is also capable of determining whether or not a digital IC is functioning correctly. It is this final feature that makes the CMOS Logic Tester a valuable addition to the "toolbox" of every digital experimenter.

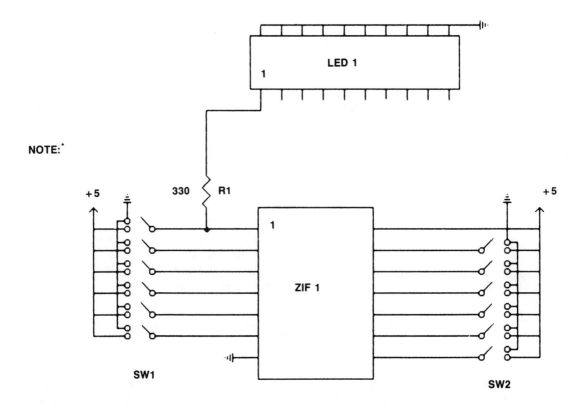

*NOTE: Repeat these connections for pins 2-6 and 8-13 of ZIF 1.

Fig. 1-54. Schematic diagram for CMOS Logic Tester.

The construction of the CMOS Logic Tester is extremely simple with the schematic diagram in Fig. 1-54. Don't underestimate the value of using a ZIF (zero insertion force) socket in this project. The use of a ZIF socket is an important construction consideration in this project.

The final circuit should easily fit inside a small plastic experimenter's box. A small housing for the CMOS Logic Tester will ensure the quick testing of numerous digital ICs during the construction of the subsequent projects within this book.

Testing the completed CMOS Logic Tester will require a sampling of various 14-pin digital ICs and a set of manufacturer's logic diagrams for each of the tested chips. Begin these performance tests, by inserting the digital gate into the ZIF socket (ZIF 1). Now begin applying both logic 1s (+ 5 V) and logic 0s (GND) to the various pins of the target IC. Compare the results with the manufacturer's logic diagrams. Once you have recorded the results for an AND, OR, and NOT gate, create your own logic diagram for an unknown digital IC. Mastering these initial steps during these testing stages will make the CMOS Logic Tester a valuable ally during future project construction.

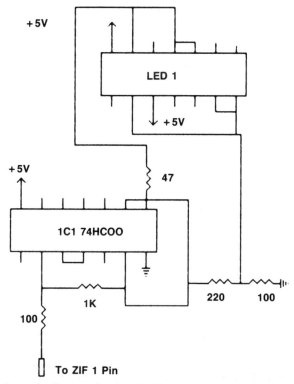

Fig. 1-55. Schematic diagram for the CMOS Logic Tester alternate logic indicator.

While the original CMOS Logic Tester is a suitable tool for determining the logic of a given digital IC's pins, there is a certain lacking in its presentation of logic. Figure 1-55 provides an exciting enhancement to the basic CMOS Logic Tester design that will correctly indicate either a logic 1 or a logic 0 state. This elaborate indicator must be interfaced to *every* test pin of the target IC. In all, this will result in 12 different 7-segment common anode displays (one display each, for every test pin of the ZIF socket). Keep this design element in mind prior to attempting this conversion.

After this modification has been installed on the CMOS Logic Tester, repeat the same series of tests that were performed on the basic CMOS Logic Tester configuration. Contrary to the "high-only" indication that is displayed with the basic design, this enhanced CMOS Logic Tester will display the exact logic for every pin of the target IC. The extra expense that is entailed with this modification can be justified based solely on the merit of this informative display.

2

Gates

CD4000 CMOS NOR

The CD4000 CMOS NOR gate is a completely I/O buffered, dual three-input plus inverter, internal logic device. Supplementary configurations of the CD4000 feature unbuffered I/O ports.

Package Configuration: 14-lead DIP

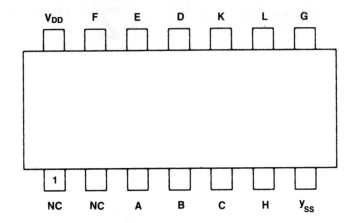

Fig. 2-1. Pin assignments for CD4000 CMOS NOR Gate.

Maximum Input Capacitance: 7.5 pF

Maximum Propagation Delay Time: 250 ns at V_{DD} = 5.0 V
120 ns at V_{DD} = 10.0 V
90 ns at V_{DD} = 15.0 V

Maximum Transition Time: 200 ns at V_{DD} = 5.0 V
100 ns at V_{DD} = 10.0 V
80 ns at V_{DD} = 15.0 V

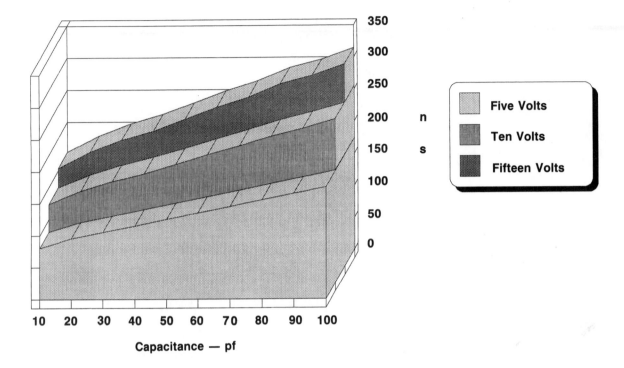

Fig. 2-2. Propagation delay for CD4000.

CD4001   CMOS NOR

The CD4001 CMOS NOR gate is a completely I/O buffered, quad dual-input, internal logic device. Supplementary configurations of the CD4001 feature unbuffered I/O ports.

Package Configuration: 14-lead DIP

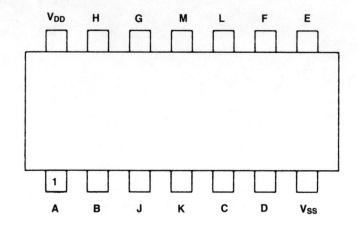

Fig. 2-3. Pin assignments for CD4001 CMOS NOR Gate.

Maximum Input Capacitance: 7.5 pF

Maximum Propagation Delay Time: 250 ns at V_{DD} = 5.0 V
120 ns at V_{DD} = 10.0 V
90 ns at V_{DD} = 15.0 V

Maximum Transition Time: 200 ns at V_{DD} = 5.0 V
100 ns at V_{DD} = 10.0 V
80 ns at V_{DD} = 15.0 V

CD4002 CMOS NOR

The CD4002 CMOS NOR gate is a completely I/O buffered, dual four-input, internal logic device. Supplementary configurations of the CD4002 feature unbuffered I/O ports.

Package Configuration: 14-lead DIP

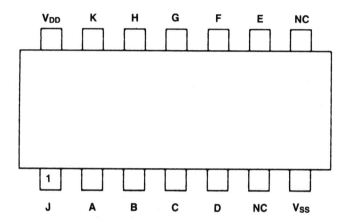

Fig. 2-4. Pin assignments for CD4002 CMOS NOR Gate.

Maximum Input Capacitance: 7.5 pF

Maximum Propagation Delay Time: 250 ns at V_{DD} = 5.0 V
120 ns at V_{DD} = 10.0 V
90 ns at V_{DD} = 15.0 V

Maximum Transition Time: 200 ns at V_{DD} = 5.0 V
100 ns at V_{DD} = 10.0 V
80 ns at V_{DD} = 15.0 V

CD4011 CMOS NAND

The CD4011 CMOS NAND gate is a completely I/O buffered, quad two-input, internal logic device. Supplementary configurations of the CD4011 feature unbuffered I/O ports.

Package Configuration: 14-lead DIP

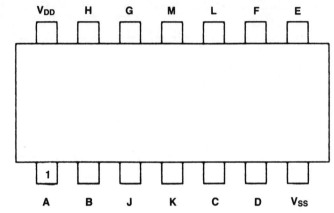

Fig. 2-5. Pin assignments for CD4011 CMOS NAND Gate.

Maximum Input Capacitance: 7.5 pF

Maximum Propagation Delay Time: 250 ns at $V_{DD} = 5.0$ V
120 ns at $V_{DD} = 10.0$ V
90 ns at $V_{DD} = 15.0$ V

Maximum Transition Time: 200 ns at $V_{DD} = 5.0$ V
100 ns at $V_{DD} = 10.0$ V
80 ns at $V_{DD} = 15.0$ V

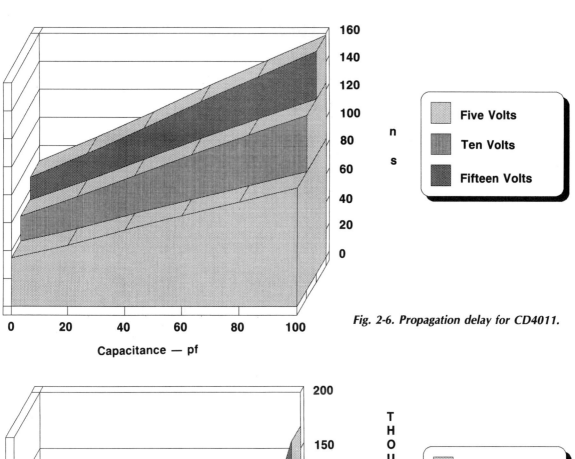

Capacitance — pf

Fig. 2-6. Propagation delay for CD4011.

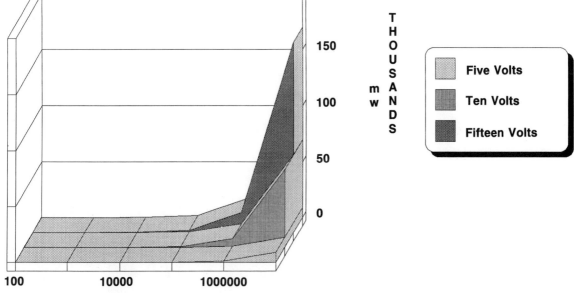

Input Frequency — Hz

Fig. 2-7. Transition time for CD4011.

CD4012 CMOS NAND

The CD4012 CMOS NAND gate is a completely I/O buffered, dual four-input, internal logic device. Supplementary configurations of the CD4012 feature unbuffered I/O ports.

Package Configuration: 14-lead DIP

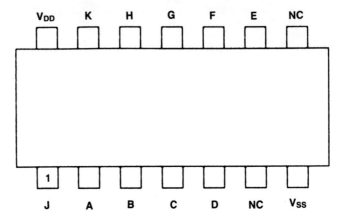

Fig. 2-8. Pin assignments for CD4012 CMOS NAND Gate.

Maximum Input Capacitance: 7.5 pF

Maximum Propagation Delay Time: 250 ns at V_{DD} = 5.0 V
120 ns at V_{DD} = 10.0 V
90 ns at V_{DD} = 15.0 V

Maximum Transition Time: 200 ns at V_{DD} = 5.0 V
100 ns at V_{DD} = 10.0 V
80 ns at V_{DD} = 15.0 V

CD4019 CMOS QUAD AND/OR

The CD4019 CMOS Quad AND/OR select gate is arranged as four select gates driving two two-input AND gates connected to a single two-input OR gate. Two control bits, Ka and Kb, are used for gate selection. If these two control bits are selected simultaneously, then an A+B logic function is implemented.

Package Configuration: 16-lead DIP

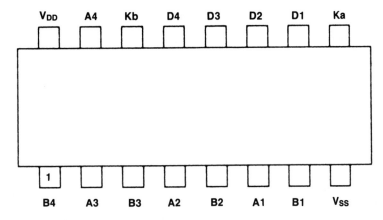

Fig. 2-9. Pin assignments for CD4019 CMOS Quad AND/OR Select Gate.

Maximum Input Capacitance: A and B Inputs = 7.5 pF
Control Ka and Kb Inputs = 15 pF

Maximum Propagation Delay Time: 300 ns at V_{DD} = 5.0 V
120 ns at V_{DD} = 10.0 V
100 ns at V_{DD} = 15.0 V

Maximum Transition Time: 200 ns at V_{DD} = 5.0 V
100 ns at V_{DD} = 10.0 V
80 ns at V_{DD} = 15.0 V

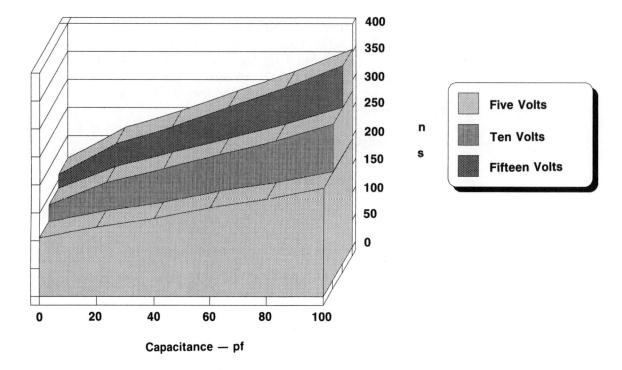

Fig. 2-10. Propagation delay for CD4019.

CD4023 CMOS NAND

The CD4023 CMOS NAND gate is a completely I/O buffered, triple three-input, internal logic device. Supplementary configurations of the CD4023 feature unbuffered I/O ports.

Package Configuration: 14-lead DIP

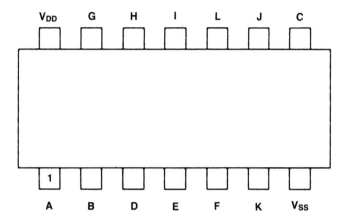

Fig. 2-11. Pin assignments for CD4023 CMOS NAND Gate.

Maximum Input Capacitance: 7.5 pF

Maximum Propagation Delay Time: 250 ns at V_{DD} = 5.0 V
120 ns at V_{DD} = 10.0 V
90 ns at V_{DD} = 15.0 V

Maximum Transition Time: 200 ns at V_{DD} = 5.0 V
100 ns at V_{DD} = 10.0 V
80 ns at V_{DD} = 15.0 V

CD4025 CMOS NOR

The CD4025 CMOS NOR gate is a completely I/O buffered, triple three-input, internal logic device. Supplementary configurations of the CD4025 feature unbuffered I/O ports.

Package Configuration: 14-lead DIP

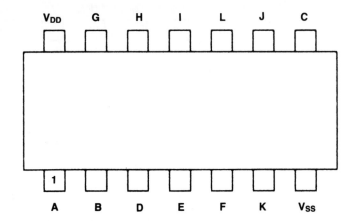

Fig. 2-12. Pin assignments for CD4025 CMOS NOR Gate.

Maximum Input Capacitance: 7.5 pF

Maximum Propagation Delay Time: 250 ns at V_{DD} = 5.0 V
120 ns at V_{DD} = 10.0 V
90 ns at V_{DD} = 15.0 V

Maximum Transition Time: 200 ns at V_{DD} = 5.0 V
100 ns at V_{DD} = 10.0 V
80 ns at V_{DD} = 15.0 V

CD4030 CMOS QUAD EXCLUSIVE-OR

The CD4030 CMOS Quad Exclusive-OR gate's internal logic is arranged as four separate exclusive-OR gates.

Package Configuration: 14-lead DIP

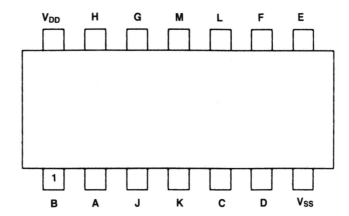

Fig. 2-13. Pin assignments for CD4030 CMOS Quad Exclusive-OR Gate.

Maximum Input Capacitance: 7.5 pF

Maximum Propagation Delay Time: 280 ns at V_{DD} = 5.0 V
130 ns at V_{DD} = 10.0 V
100 ns at V_{DD} = 15.0 V

Maximum Transition Time: 200 ns at V_{DD} = 5.0 V
100 ns at V_{DD} = 10.0 V
80 ns at V_{DD} = 15.0 V

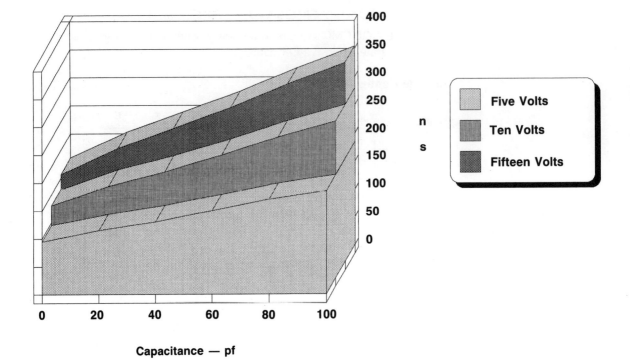

Fig. 2-14. Propagation delay for CD4030.

CD4048 CMOS Multifunction Expandable 8-Input Gate

The CD4048 CMOS Multifunction Expandable 8-Input Gate is manipulated through four control inputs, Ka, Kb, Kc, and Kd. The first three inputs, Ka, Kb, and Kc, are two-state binary controls that are used for selecting eight different gate functions: OR, AND, NOR, NAND, OR/AND, OR/NAND, AND/OR, and AND/NOR.

The final control input, Kd, generates a three-state output. Therefore, an internally determined 1 or 0 logic output is possible when Kd is high, and an open circuit output is produced for a low condition on Kd.

This CMOS device also has the ability to add extra input lines through the EXPAND input. In this arrangement, additional CD4048 ICs can be cascaded together for multiples-of-eight multifunction gates. A design notation for the EXPAND input requires that it be connected to Vss when not used in a circuit.

Package Configuration: 16-lead DIP

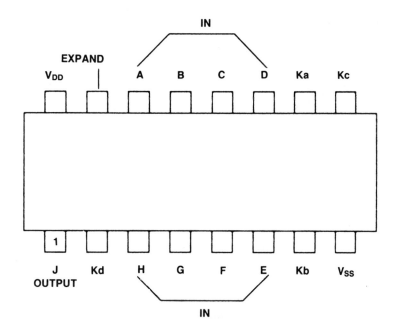

Fig. 2-15. Pin assignments for CD4048 CMOS Multifunction Expandable 8-Input Gate.

Maximum Input Capacitance: 7 pF

Maximum Three-State Output Capacitance: 10 pF

Maximum Ka-to-Output Propagation Delay Time:
600 ns at V_{DD} = 5.0 V
300 ns at V_{DD} = 10.0 V
240 ns at V_{DD} = 15.0 V

Maximum Kb-to-Output Propagation Delay Time:
450 ns at V_{DD} = 5.0 V
170 ns at V_{DD} = 10.0 V
110 ns at V_{DD} = 15.0 V

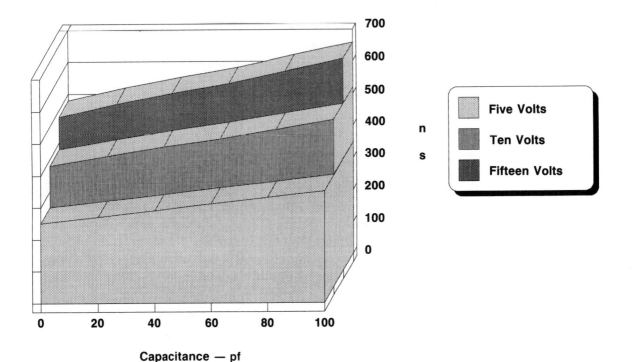

Fig. 2-16. Propagation delay for CD4048.

Maximum Kc-to-Output Propagation Delay Time:
280 ns at V_{DD} = 5.0 V
100 ns at V_{DD} = 10.0 V
80 ns at V_{DD} = 15.0 V

Maximum EXPAND-to-Output Propagation Delay Time:
380 ns at V_{DD} = 5.0 V
180 ns at V_{DD} = 10.0 V
130 ns at V_{DD} = 15.0 V

Maximum Three-State Propagation Delay Time:
160 ns at $V_{DD} = 5.0$ V
70 ns at $V_{DD} = 10.0$ V
50 ns at $V_{DD} = 15.0$ V

Maximum Transition Time: 200 ns at $V_{DD} = 5.0$ V
100 ns at $V_{DD} = 10.0$ V
80 ns at $V_{DD} = 15.0$ V

CD4068 CMOS EIGHT-INPUT NAND/AND

The CD4068 CMOS Eight-Input NAND/AND gate's internal logic is arranged as a buffered, 8-input, positive-logic, NAND and AND gate.

Package Configuration: 14-lead DIP

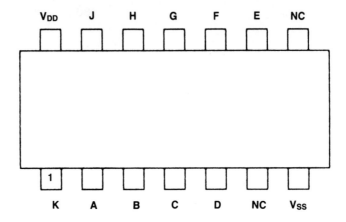

Fig. 2-17. Pin assignments for CD4068 CMOS Eight-Input NAND/AND Gate.

Maximum Input Capacitance: 7.5 pF

Maximum Propagation Delay Time: 300 ns at V_{DD} = 5.0 V
150 ns at V_{DD} = 10.0 V
110 ns at V_{DD} = 15.0 V

Maximum Transition Time: 200 ns at V_{DD} = 5.0 V
100 ns at V_{DD} = 10.0 V
80 ns at V_{DD} = 15.0 V

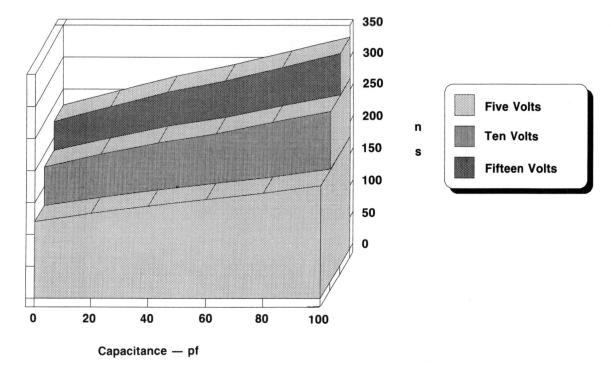

Capacitance — pf

Fig. 2-18. Propagation delay for CD4068.

CD4070 CMOS QUAD EXCLUSIVE-OR

The CD4070 CMOS Quad Exclusive-OR gate's internal logic is arranged as four separate exclusive-OR gates.

Package Configuration: 14-lead DIP

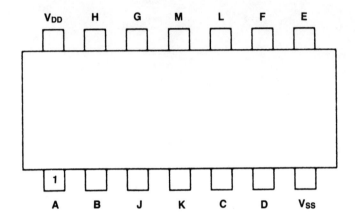

Fig. 2-19. Pin assignments for CD4070 CMOS Quad Exclusive-OR Gate.

Maximum Input Capacitance: 7.5 pF

Maximum Propagation Delay Time: 280 ns at V_{DD} = 5.0 V
130 ns at V_{DD} = 10.0 V
100 ns at V_{DD} = 15.0 V

Maximum Transition Time: 200 ns at V_{DD} = 5.0 V
100 ns at V_{DD} = 10.0 V
80 ns at V_{DD} = 15.0 V

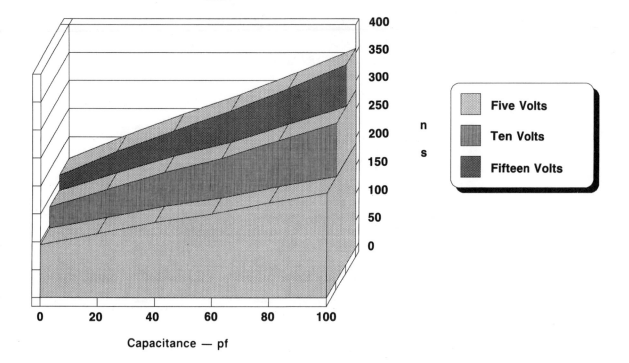

Fig. 2-20. Propagation delay for CD4070.

CD4071 CMOS QUAD TWO-INPUT OR

The CD4071 CMOS Quad Two-Input OR gate's internal logic is arranged as four, two-input, positive-logic, OR gates.

Package Configuration: 14-lead DIP

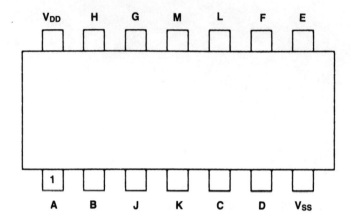

Fig. 2-21. Pin assignments for CD4071 CMOS Quad Two-Input OR Gate.

Maximum Input Capacitance: 7.5 pF

Maximum Propagation Delay Time: 250 ns at V_{DD} = 5.0 V
120 ns at V_{DD} = 10.0 V
90 ns at V_{DD} = 15.0 V

Maximum Transition Time: 200 ns at V_{DD} = 5.0 V
100 ns at V_{DD} = 10.0 V
80 ns at V_{DD} = 15.0 V

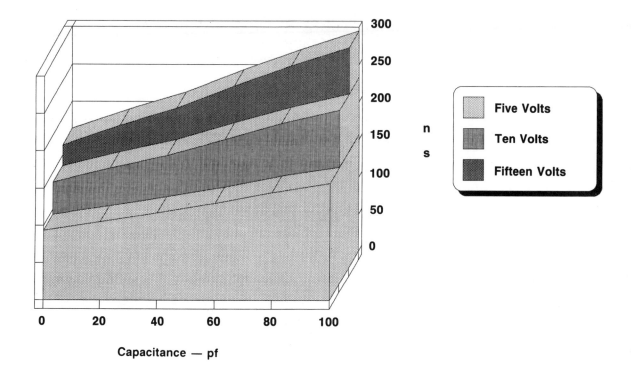

Fig. 2-22. Propagation delay for CD4071.

CD4072 CMOS DUAL FOUR-INPUT OR

The CD4072 CMOS Dual Four-Input OR gate's internal logic is arranged as two, four-input, positive-logic, OR gates.

Package Configuration: 14-lead DIP

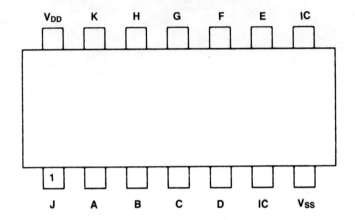

Fig. 2-23. Pin assignments for CD4072 CMOS Dual Four-Input OR Gate.

Maximum Input Capacitance: 7.5 pF

Maximum Propagation Delay Time: 250 ns at V_{DD} = 5.0 V
120 ns at V_{DD} = 10.0 V
90 ns at V_{DD} = 15.0 V

Maximum Transition Time: 200 ns at V_{DD} = 5.0 V
100 ns at V_{DD} = 10.0 V
80 ns at V_{DD} = 15.0 V

CD4073 CMOS TRIPLE THREE-INPUT AND

The CD4073 CMOS Triple Three-Input AND gate has an internal logic arranged as three, three-input, AND gates.

Package Configuration: 14-lead DIP

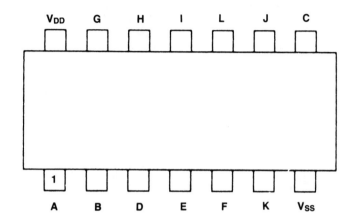

Fig. 2-24. Pin assignments for CD4073 CMOS Triple Three-Input AND Gate.

Maximum Input Capacitance: 7.5 pF

Maximum Propagation Delay Time: 250 ns at V_{DD} = 5.0 V
120 ns at V_{DD} = 10.0 V
90 ns at V_{DD} = 15.0 V

Maximum Transition Time: 200 ns at V_{DD} = 5.0 V
100 ns at V_{DD} = 10.0 V
80 ns at V_{DD} = 15.0 V

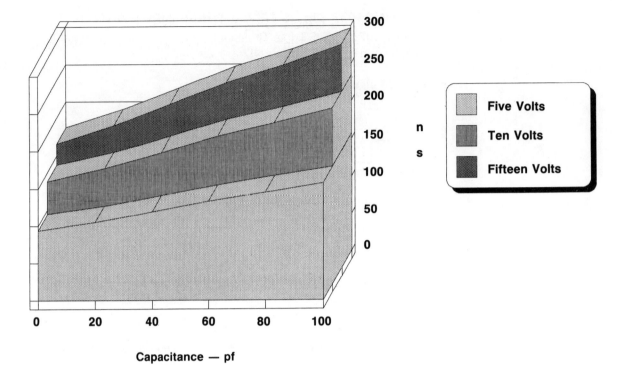

Fig. 2-25. Propagation delay for CD4073.

CD4075 CMOS TRIPLE THREE-INPUT OR

The CD4075 CMOS Triple Three-Input OR gate has an internal logic arranged as three, three-input, positive-logic, OR gates.

Package Configuration: 14-lead DIP

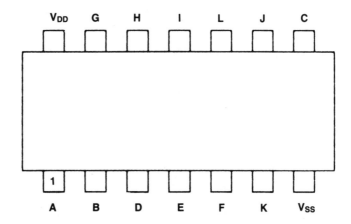

Fig. 2-26. Pin assignments for CD4075 CMOS Triple Three-Input OR Gate.

Maximum Input Capacitance: 7.5 pF

Maximum Propagation Delay Time: 250 ns at V_{DD} = 5.0 V
120 ns at V_{DD} = 10.0 V
90 ns at V_{DD} = 15.0 V

Maximum Transition Time: 200 ns at V_{DD} = 5.0 V
100 ns at V_{DD} = 10.0 V
80 ns at V_{DD} = 15.0 V

CD4077 CMOS QUAD EXCLUSIVE-NOR

The CD4077 CMOS Quad Exclusive-NOR gate's internal logic is arranged as four separate exclusive-NOR gates.

Package Configuration: 14-lead DIP

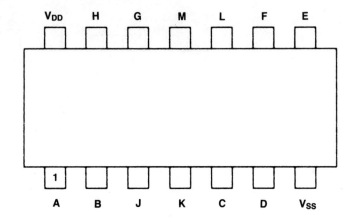

Fig. 2-27. Pin assignments for CD4077 CMOS Quad Exclusive-NOR Gate.

Maximum Input Capacitance: 7.5 pF

Maximum Propagation Delay Time: 280 ns at V_{DD} = 5.0 V
130 ns at V_{DD} = 10.0 V
100 ns at V_{DD} = 15.0 V

Maximum Transition Time: 200 ns at V_{DD} = 5.0 V
100 ns at V_{DD} = 10.0 V
80 ns at V_{DD} = 15.0 V

CD4078 CMOS EIGHT-INPUT NOR/OR

The CD4078 CMOS Eight-Input NOR/OR gate's internal logic is arranged as a single buffered, positive-logic, 8-input, NOR and OR gate.

Package Configuration: 14-lead DIP

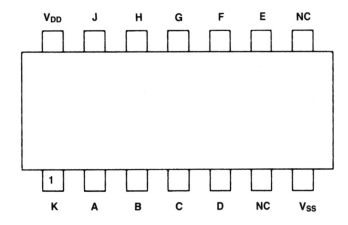

Fig. 2-28. Pin assignments for CD4078 CMOS Eight-Input NOR/OR Gate.

Maximum Input Capacitance: 7.5 pF

Maximum Propagation Delay Time: 300 ns at V_{DD} = 5.0 V
150 ns at V_{DD} = 10.0 V
110 ns at V_{DD} = 15.0 V

Maximum Transition Time: 200 ns at V_{DD} = 5.0 V
100 ns at V_{DD} = 10.0 V
80 ns at V_{DD} = 15.0 V

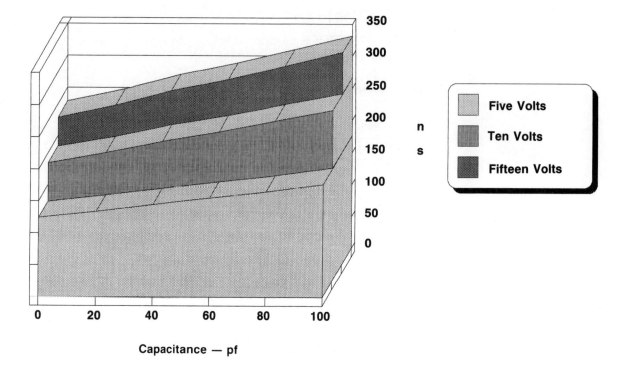

Fig. 2-29. Propagation Delay for CD4078.

CD4081 CMOS QUAD TWO-INPUT AND

The CD4081 CMOS Quad Two-Input AND gate's internal logic is arranged as four, two-input AND gates.

Package Configuration: 14-lead DIP

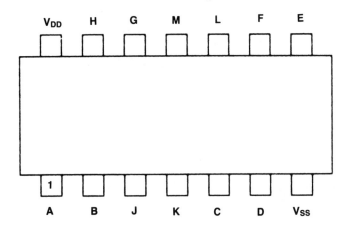

Fig. 2-30. Pin assignments for CD4081 CMOS Quad Two-Input AND Gate.

Maximum Input Capacitance: 7.5 pF

Maximum Propagation Delay Time: 250 ns at V_{DD} = 5.0 V
120 ns at V_{DD} = 10.0 V
90 ns at V_{DD} = 15.0 V

Maximum Transition Time: 200 ns at V_{DD} = 5.0 V
100 ns at V_{DD} = 10.0 V
80 ns at V_{DD} = 15.0 V

CD4082 CMOS DUAL FOUR-INPUT AND

The CD4082 CMOS Dual Four-Input AND gate's internal logic is arranged as two, four-input AND gates.

Package Configuration: 14-lead DIP

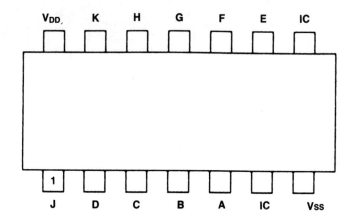

Fig. 2-31. Pin assignments for CD4082 CMOS Dual Four-Input AND Gate.

Maximum Input Capacitance: 7.5 pF

Maximum Propagation Delay Time: 250 ns at V_{DD} = 5.0 V
120 ns at V_{DD} = 10.0 V
90 ns at V_{DD} = 15.0 V

Maximum Transition Time: 200 ns at V_{DD} = 5.0 V
100 ns at V_{DD} = 10.0 V
80 ns at V_{DD} = 15.0 V

CD4086 CMOS EXPANDABLE
FOUR-WIDE TWO-INPUT AND-OR-INVERT

The CD4086 CMOS Expandable Four-Wide Two-Input AND-OR-INVERT gate is arranged in an internal logic of one 4-wide, 2-input AND- OR-INVERT gate with buffered outputs. Control over the logic of the CD4086 is provided with the INHIBIT and ENABLE inputs. In order to enable the 4-wide AND-OR-INVERT function, the INHIBIT input must be connected to Vss and the ENABLE input must be tied to VDD.

Another input, J, is used to expand the width of the CD4086 in multiples of four. In cascading two CD4086 ICs, for example, the J input from the first gate is tied to the ENABLE pin of the second. Likewise, additional stages to this expansion can be made through further connections between the J/ENABLE pins.

Package Configuration: 14-lead DIP

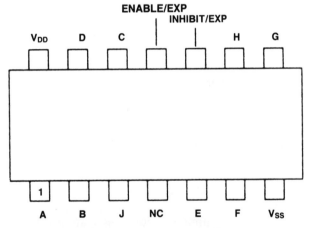

Fig. 2-32. Pin assignments for CD4086 CMOS Expandable Four-Wide Two-Input AND-OR-INVERT Gate.

Maximum Input Capacitance: 7.5 pF

Maximum Data Propagation Delay Time; High-to-Low:
450 ns at VDD = 5.0 V
180 ns at VDD = 10.0 V
120 ns at VDD = 15.0 V

Maximum Data Propagation Delay Time; High-to-Low:
450 ns at VDD = 5.0 V
180 ns at VDD = 10.0 V
120 ns at VDD = 15.0 V

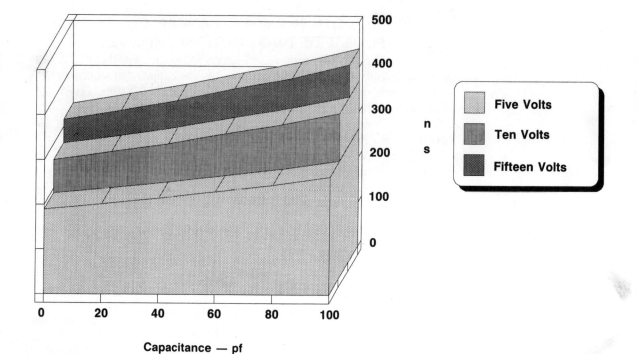

Fig. 2-33. Propagation delay for CD4086.

Maximum Data Propagation Delay Time; Low-to-High:
620 ns at V_{DD} = 5.0 V
250 ns at V_{DD} = 10.0 V
180 ns at V_{DD} = 15.0 V

Maximum Inhibit Propagation Delay Time; High-to-Low:
300 ns at V_{DD} = 5.0 V
120 ns at V_{DD} = 10.0 V
80 ns at V_{DD} = 15.0 V

Maximum Inhibit Propagation Delay Time; Low-to-High:
500 ns at V_{DD} = 5.0 V
200 ns at V_{DD} = 10.0 V
140 ns at V_{DD} = 15.0 V

Maximum Transition Time:
200 ns at V_{DD} = 5.0 V
100 ns at V_{DD} = 10.0 V
80 ns at V_{DD} = 15.0 V

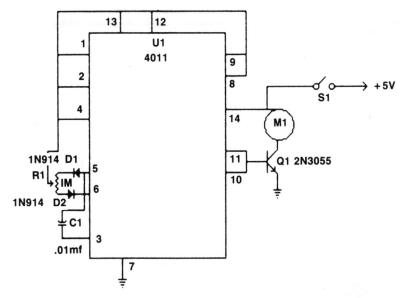

Fig. 2-34. Schematic diagram for Motor Speed Control.

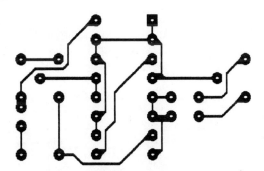

Fig. 2-35. Solder side of the Motor Speed Control PCB template.

MOTOR SPEED CONTROL

An interesting application for a CMOS 4011 NAND gate IC is in controlling the apparent speed of a small electric motor (see Fig. 2-34). This variable speed is supplied as a series of pulses, as the number of power pulses decreases, the working speed of the motor decreases. Conversely, as the number of pulses increases, the operating speed of the motor increases.

Potentiometer R1 is used for varying the speed of the connected electric motor (see Figs. 2-35 and 2-36). Virtually any electric dc motor, up to 5-volts,

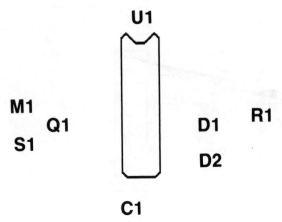

Fig. 2-36. Parts layout for the Motor Speed Control PCB.

can be controlled with this project. Ideally, binding posts should be soldered onto the final PCB at the motor M1 attachment points. This construction procedure would permit any electric motor to be quickly interfaced, and subsequently, governed by this project.

METAL DETECTOR

Another CMOS gate IC that can be applied to discrete utility devices is the CD4030 Quad exclusive-OR Gate. In this case, each of the gates are

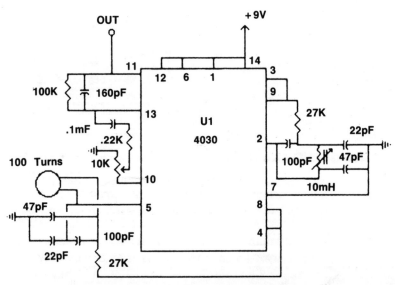

Fig. 2-37. Schematic diagram for Metal Detector.

configured as an oscillator. Three of these oscillators are then arranged as frequency, gain, and volume controls for the fourth oscillator. This last exclusive-ORed oscillator serves as a locating coil in the Metal Detector. A small wire coil is used as the inductance element for this final CD4030 oscillator (see Fig. 2-37).

When powered by a single 9-volt battery, the Metal Detector has a sensitivity of several inches. By adjusting the tuning coil, an exact frequency separation can be produced that will enhance the inductance change generated from the locating coil oscillator. Use the adjustable potentiometer for controlling the volume of the detection signal from the Metal Detector's output.

3

Buffers and Inverters

CD4007 CMOS DUAL
COMPLEMENTARY PAIR PLUS INVERTER

The CD4007 CMOS Dual Complementary Pair Plus Inverter consists of three n-channel enhancement-mode MOS transistors (E-MOS), and three p-channel E-MOS transistors. Numerous logic configurations can be designed through the selective arrangement of the source, drain, and gate terminals of these transistors.

Package Configuration: 14-lead DIP

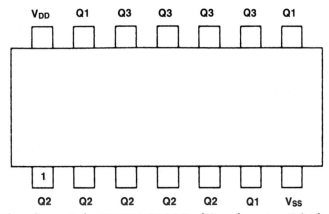

Fig. 3-1. Pin assignments for CD4007 CMOS Dual Complementary Pair Plus Inverter.

Maximum Input Capacitance: 15 pF

Maximum Propagation Delay Time: 110 ns at V_{DD} = 5.0 V
60 ns at V_{DD} = 10.0 V
50 ns at V_{DD} = 15.0 V

Maximum Transition Time: 200 ns at V_{DD} = 5.0 V
100 ns at V_{DD} = 10.0 V
80 ns at V_{DD} = 15.0 V

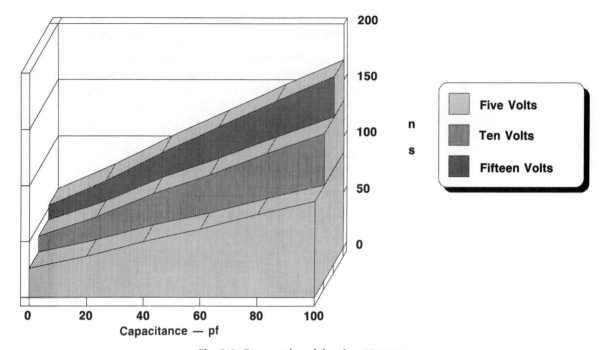

Fig. 3-2. Propagation delay for CD4007.

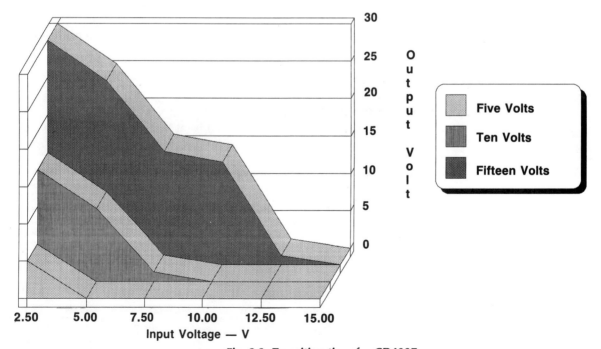

Fig. 3-3. Transition time for CD4007.

CD4009 CMOS HEX INVERTING BUFFER/CONVERTER

The CD4009 CMOS Hex Inverting Buffer/Converter has an internal logic arranged as six inverting logic converters. These inverters can be used as either interface converters between CMOS logic, DTL, and TTL logic devices or as CMOS high-sink current drivers.

Package Configuration: 16-lead DIP

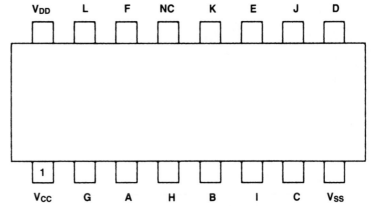

Fig. 3-4. Pin assignments for CD4009 CMOS Hex Inverting Buffer/Converter.

Maximum Input Capacitance: 22.5 pF

Maximum Propagation Delay Time; High-to-Low:
60 ns at V_{DD} = 5.0 V
40 ns at V_{DD} = 10.0 V
30 ns at V_{DD} = 15.0 V

Maximum Propagation Delay Time; Low-to-High:
140 ns at V_{DD} = 5.0 V
80 ns at V_{DD} = 10.0 V
60 ns at V_{DD} = 15.0 V

Maximum Transition Time; High-to-Low:
70 ns at V_{DD} = 5.0 V
40 ns at V_{DD} = 10.0 V
30 ns at V_{DD} = 15.0 V

Maximum Transition Time; Low-to-High:
350 ns at V_{DD} = 5.0 V
150 ns at V_{DD} = 10.0 V
110 ns at V_{DD} = 15.0 V

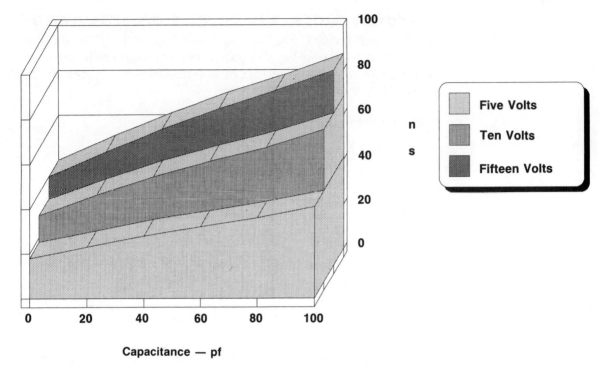

Capacitance — pf

Fig. 3-5. Propagation delay for CD4009.

CD4010 CMOS HEX NON-INVERTING BUFFER/CONVERTER

The CD4010 CMOS Hex Non-Inverting Buffer/Converter has an internal logic arranged as six non-inverting logic converters. These buffers can be used as either interface converters between CMOS logic, DTL, and TTL logic devices or as CMOS high-sink current drivers.

Package Configuration: 16-lead DIP

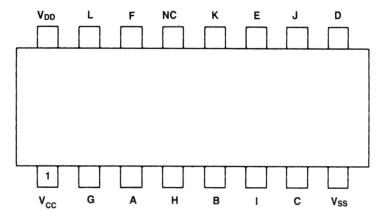

Fig. 3-6. Pin assignments for CD4010 CMOS Hex Non-Inverting Buffer/Converter.

Maximum Input Capacitance: 7.5 pF

Maximum Propagation Delay Time; High-to-Low:
130 ns at V_{DD} = 5.0 V
70 ns at V_{DD} = 10.0 V
50 ns at V_{DD} = 15.0 V

Maximum Propagation Delay Time; Low-to-High:
200 ns at V_{DD} = 5.0 V
100 ns at V_{DD} = 10.0 V
70 ns at V_{DD} = 15.0 V

Maximum Transition Time; High-to-Low:
70 ns at V_{DD} = 5.0 V
40 ns at V_{DD} = 10.0 V
30 ns at V_{DD} = 15.0 V

Maximum Transition Time; Low-to-High:
350 ns at V_{DD} = 5.0 V
150 ns at V_{DD} = 10.0 V
110 ns at V_{DD} = 15.0 V

CD4041 CMOS QUAD TRUE/COMPLEMENT BUFFER

The CD4041 CMOS Quad True/Complement Buffer is arranged with an internal logic of four true/complement, n-channel and p-channel buffers. Each of these buffers demonstrates a high current sourcing, and a high current sinking attribute. These features make the CD4041 an excellent, low power dissipation buffer, line driver, and TTL logic driver.

Package Configuration: 14-lead DIP

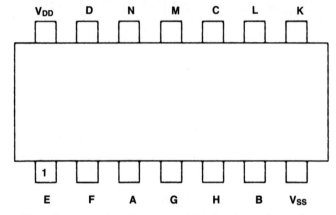

Fig. 3-7. Pin assignments for CD4041 CMOS Quad True/Complement Buffer.

Maximum Input Capacitance: 22.5 pF

Maximum Propagation Delay Time: 120 ns at V_{DD} = 5.0 V
70 ns at V_{DD} = 10.0 V
50 ns at V_{DD} = 15.0 V

Maximum Transition Time: 80 ns at V_{DD} = 5.0 V
40 ns at V_{DD} = 10.0 V
30 ns at V_{DD} = 15.0 V

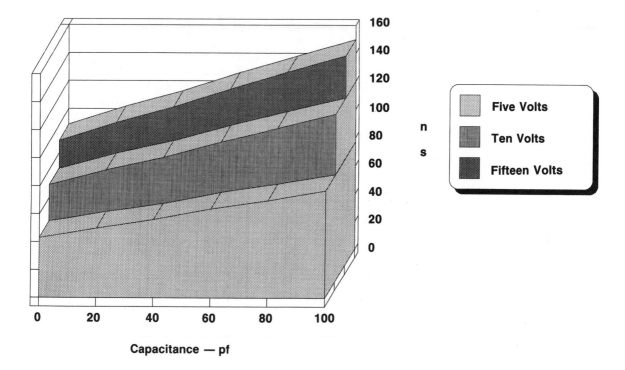

Fig. 3-8. Propagation delay for CD4041.

CD4049 CMOS INVERTING HEX BUFFER/CONVERTER

The CD4049 CMOS Inverting Hex Buffer/Converter has an internal logic of six, single voltage supply, logic inverting buffers. The CD4049 is capable of converting DTL and TTL devices to CMOS logic levels. In this application, a single CD4049 is able to drive two DTL or TTL loads. Based on its single power supply requirement, CD4049 is an enhanced replacement for CD4009.

Package Configuration: 16-lead DIP

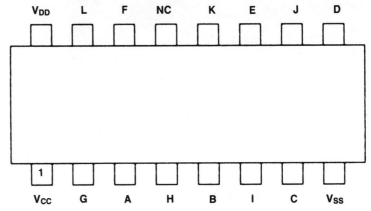

Fig. 3-9. Pin assignments for CD4049 CMOS Inverting Hex Buffer/Converter.

Maximum Input Capacitance: 22.5 pF

Maximum Propagation Delay Time; High-to-Low:
   65 ns at V_{DD} = 5.0 V
   40 ns at V_{DD} = 10.0 V
   30 ns at V_{DD} = 15.0 V

Maximum Propagation Delay Time; Low-to-High:
   120 ns at V_{DD} = 5.0 V
   65 ns at V_{DD} = 10.0 V
   50 ns at V_{DD} = 15.0 V

Maximum Transition Time; High-to-Low:
   60 ns at V_{DD} = 5.0 V
   40 ns at V_{DD} = 10.0 V
   30 ns at V_{DD} = 15.0 V

Maximum Transition Time; Low-to-High:
   160 ns at V_{DD} = 5.0 V
   80 ns at V_{DD} = 10.0 V
   60 ns at V_{DD} = 15.0 V

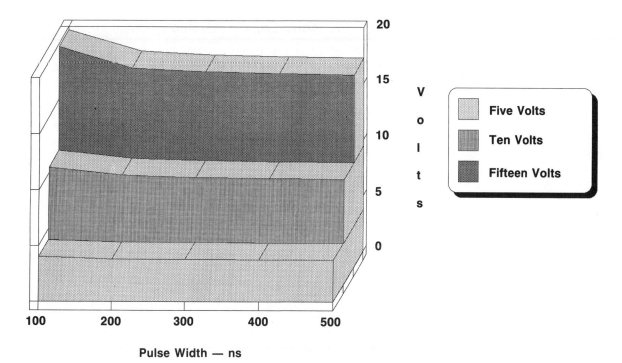

Pulse Width — ns

Fig. 3-10. Propagation delay for CD4049.

CD4069 CMOS HEX INVERTER

The CD4069 CMOS Hex Inverter has an internal logic of six inverters. This general-purpose logic inverter lacks the power- handling DTL/TTL-driver capabilities of the CD4009 and the CD4049.

Package Configuration: 14-lead DIP

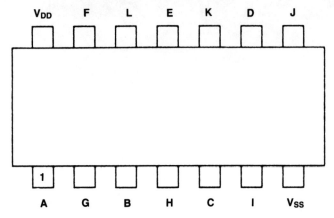

Fig. 3-11. Pin assignments for CD4069 CMOS Hex Inverter.

Maximum Input Capacitance: 15 pF

Maximum Propagation Delay Time: 110 ns at V_{DD} = 5.0 V
60 ns at V_{DD} = 10.0 V
50 ns at V_{DD} = 15.0 V

Maximum Transition Time: 200 ns at V_{DD} = 5.0 V
100 ns at V_{DD} = 10.0 V
80 ns at V_{DD} = 15.0 V

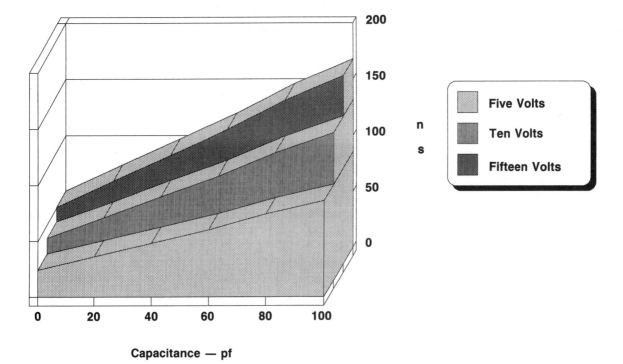

Capacitance — pf

Fig. 3-12. Propagation delay for CD4069.

CD40107 CMOS DUAL TWO-INPUT NAND BUFFER/DRIVER

The CD40107 CMOS Dual Two-Input NAND Buffer/Driver has an internal logic of two, two-input NAND buffers with open drain n-channel outputs. Other features of the CD40107 include high current sinking outputs, and a wired-OR capability.

Package Configuration: 8-lead DIP

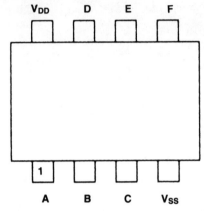

Fig. 3-13. Pin assignments for CD40107 CMOS Dual Two-Input NAND Buffer/Driver.

Maximum Propagation Delay Time; High-to-Low:
200 ns at V_{DD} = 5.0 V
90 ns at V_{DD} = 10.0 V
60 ns at V_{DD} = 15.0 V

Maximum Propagation Delay Time; Low-to-High:
200 ns at V_{DD} = 5.0 V
120 ns at V_{DD} = 10.0 V
100 ns at V_{DD} = 15.0 V

Maximum Transition Time; High-to-Low:
100 ns at V_{DD} = 5.0 V
40 ns at V_{DD} = 10.0 V
20 ns at V_{DD} = 15.0 V

Maximum Transition Time; Low-to-High:
100 ns at V_{DD} = 5.0 V
70 ns at V_{DD} = 10.0 V
50 ns at V_{DD} = 15.0 V

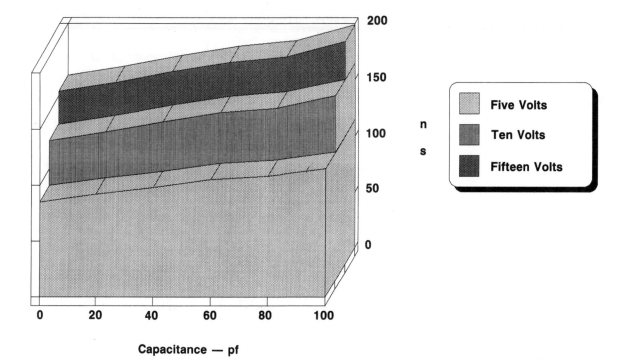

Capacitance — pf

Fig. 3-14. Propagation delay for CD40107.

CD4502 CMOS STROBED HEX INVERTER/BUFFER

The CD4502 CMOS Strobed Hex Inverter/Buffer has an internal logic of six inverters with three-state outputs. Up to two TTL loads can be driven by the CD4502. The OUTPUT DISABLE pin is used for switching the state of *all* outputs. A logic 0 on the OUTPUT DISABLE permits the INHIBIT pin to control the logic of the outputs (e.g., all outputs are a logic 0, if INHIBIT is a logic 1). Furthermore, a logic 1 on the OUTPUT DISABLE forces all of the outputs into a high-impedance state.

Package Configuration: 16-lead DIP

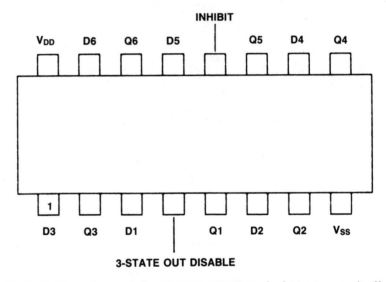

Fig. 3-15. Pin assignments for CD4502 CMOS Strobed Hex Inverter/Buffer.

Maximum Input Capacitance: 7.5 pF

Maximum Data/Inhibit Delay Time; High-to-Low:
270 ns at V_{DD} = 5.0 V
120 ns at V_{DD} = 10.0 V
80 ns at V_{DD} = 15.0 V

Maximum Data/Inhibit Delay Time; Low-to-High:
380 ns at V_{DD} = 5.0 V
180 ns at V_{DD} = 10.0 V
130 ns at V_{DD} = 15.0 V

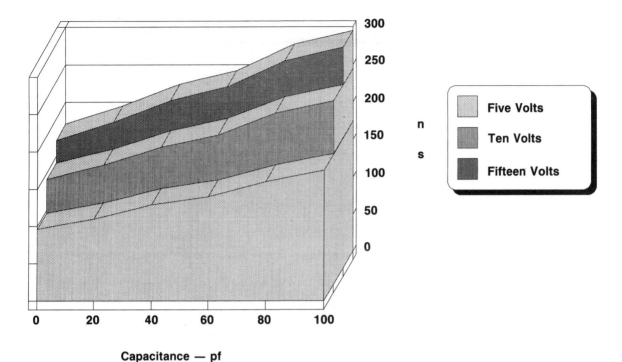

Fig. 3-16. Propagation delay for CD4502.

Maximum Transition Time; High-to-Low:
    120 ns at V_{DD} = 5.0 V
    60 ns at V_{DD} = 10.0 V
    40 ns at V_{DD} = 15.0 V

Maximum Transition Time; Low-to-High:
    200 ns at V_{DD} = 5.0 V
    100 ns at V_{DD} = 10.0 V
    80 ns at V_{DD} = 15.0 V

Maximum Disable Delay Time; High Output-to-High-Impedance:
    120 ns at V_{DD} = 5.0 V
    80 ns at V_{DD} = 10.0 V
    60 ns at V_{DD} = 15.0 V

Maximum Disable Delay Time; High-Impedance-to-High Output:
    220 ns at V_{DD} = 5.0 V
    100 ns at V_{DD} = 10.0 V
    80 ns at V_{DD} = 15.0 V

Maximum Disable Delay Time; Low Output-to-High-Impedance:

250 ns at V_{DD} = 5.0 V
130 ns at V_{DD} = 10.0 V
110 ns at V_{DD} = 15.0 V

Maximum Disable Delay Time; High-Impedance-to-Low Output:

250 ns at V_{DD} = 5.0 V
110 ns at V_{DD} = 10.0 V
80 ns at V_{DD} = 15.0 V

CD4503 CMOS THREE-STATE NON-INVERTING HEX BUFFER

The CD4503 CMOS Three-State Non-Inverting Hex Buffer is arranged with an internal logic of six, three-state non-inverting buffers. Each output has a high current sourcing, and current sinking capability. One TTL load can be driven by the CD4503. An output disable control, DISABLE A and DISABLE B is used for manipulating two separate banks of buffers. DISABLE A controls four buffers, while DISABLE B controls two buffers.

Package Configuration: 16-lead DIP

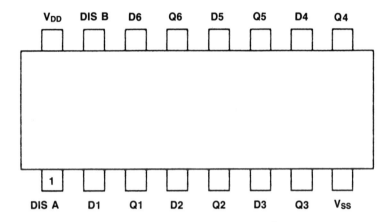

Fig. 3-17. Pin assignments for CD4503 CMOS Three-State Non-Inverting Hex Buffer.

Maximum Propagation Delay Time; High-to-Low:
110 ns at V_{DD} = 5.0 V
50 ns at V_{DD} = 10.0 V
35 ns at V_{DD} = 15.0 V

Maximum Propagation Delay Time; Low-to-High:
150 ns at V_{DD} = 5.0 V
70 ns at V_{DD} = 10.0 V
50 ns at V_{DD} = 15.0 V

Maximum Transition Time; High-to-Low:
70 ns at V_{DD} = 5.0 V
40 ns at V_{DD} = 10.0 V
25 ns at V_{DD} = 15.0 V

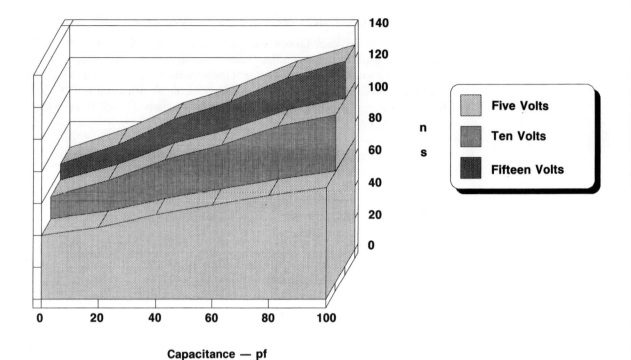

Capacitance — pf

Fig. 3-18. Propagation delay for CD4503.

Maximum Transition Time; Low-to-High:
90 ns at V_{DD} = 5.0 V
45 ns at V_{DD} = 10.0 V
35 ns at V_{DD} = 15.0 V

Maximum Propagation Delay Time; High Output-to-High-Impedance:
140 ns at V_{DD} = 5.0 V
60 ns at V_{DD} = 10.0 V
50 ns at V_{DD} = 15.0 V

[NOTE: These time specifications are the same for High-Impedance- to-High Output.]

Maximum Disable Delay Time; Low Output-to-High-Impedance:
180 ns at V_{DD} = 5.0 V
80 ns at V_{DD} = 10.0 V
70 ns at V_{DD} = 15.0 V

[NOTE: These time specifications are the same for High-Impedance- to-Low Output.]

COMPARATOR

One of the more useful microcomputer circuits is the comparator. Basically, this circuit is able to compare the logic of two or more input signals, and display the result as either a 1 or 0 logic. In a typical comparator, when the input signals are equal (i.e., A = B; in a two-signal comparator or ABCD = EFGH; in a four-signal comparator), the output result is a logic 0. Conversely, for all other possible input logic combinations, the output will be a logic 1.

Figure 3-19 is a simple four-signal or Two-Bit Comparator. In this circuit, a CD4049 Inverting Hex Buffer/Converter is used in a logic conversion role.

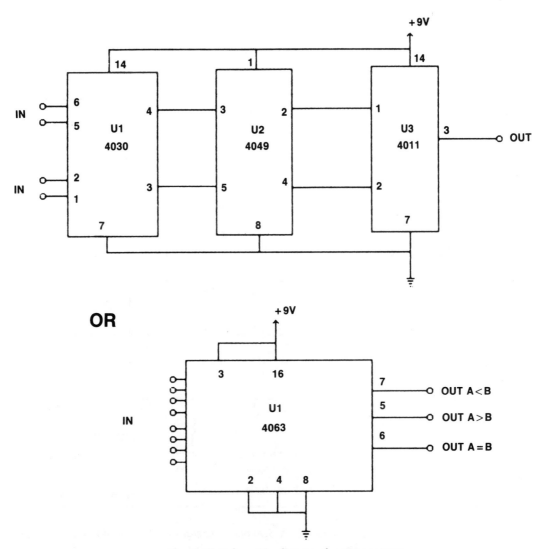

Fig. 3-19. Schematic diagram for Comparator.

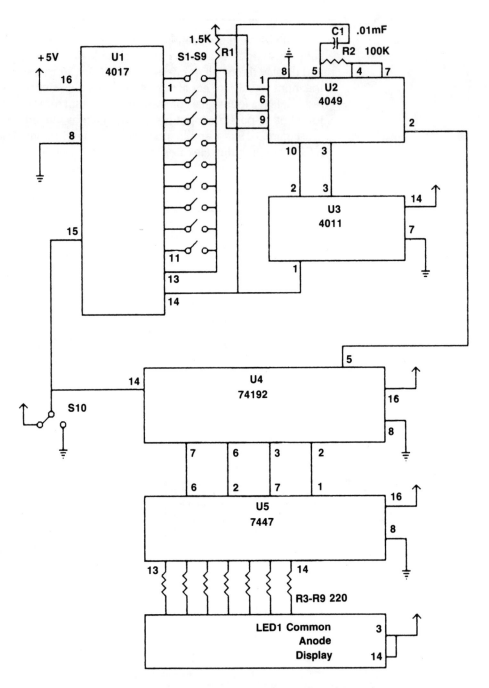

Fig. 3-20. Schematic diagram for Keyboard Encoder.

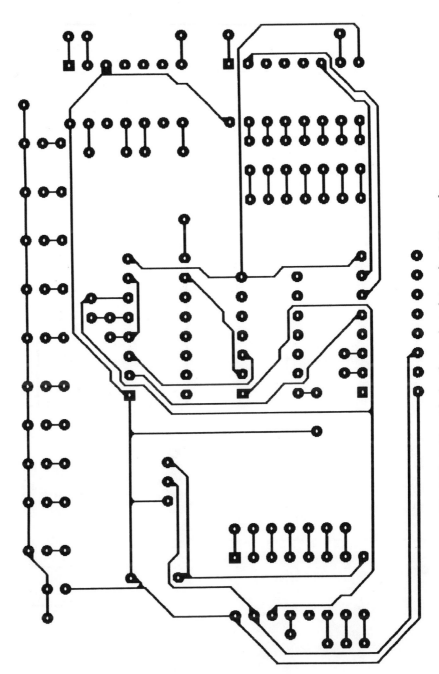

Fig. 3-21. Solder side for the Keyboard Encoder PCB template.

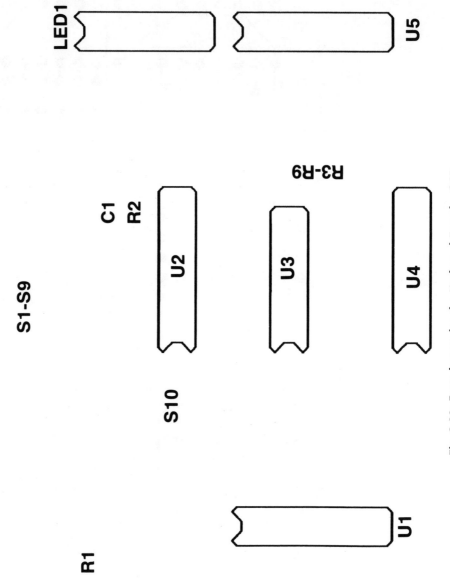

Fig. 3-22. Parts layout for the Keyboard Encoder PCB.

As presented in this schematic, two two-bit "words" are compared with each other. Only when all four of these signals are equal will the output be a logic 0. Based on this function, a simple LED display could be interfaced with the output for indicating the resultant logic from the Two-Bit Comparator.

KEYBOARD ENCODER

Another interesting project that utilizes both TTL and CMOS ICs is the keyboard encoder. A CD4049 Inverting Hex Buffer/Converter is used for providing the TTL-to-CMOS interface. Figure 3-20 contains all of the needed information for constructing a simple BCD keyboard encoder.

This ten-key encoder generates a displayed digit that corresponds to the key number that is pressed. Test the final construction by pressing a key and noting the digit that is displayed. Label each key with the appropriate number according to the results of these tests. The keyboard encoder is now ready for operation.

One of the best methods for building the keyboard encoder is with a custom-designed PCB (see Figs. 3-21 and 3-22). In this example, numerous jumpers have been used to afford the maximum construction flexibility. These jumpers can be used for elevating the attachment of the data entry switches S1-S9 and the output display LED1. For example, a hidden 9-position SPST DIP switch could be used for selecting the final output number. This method would allow a constant number to be displayed on the 7- segment display. Alternatively, a 9-key keyboard could be interfaced to this PCB with a constantly changing value displayed for each keypress. In either case, switch S10 is used for controlling the operation of the Keyboard Encoder.

Based on a single common anode 7-segment LED (Light Emitting Diode) display, this encoder can be easily expanded to represent a complete 64-key ASCII keyboard system. This expansion would require a character-generating ROM. One method for obtaining this memory IC is through a programmed EPROM. Refer to Chapter 10 for information on designing with EPROMs.

4

Counters

CD4017 CMOS DECADE COUNTER/DIVIDER

The CD4017 CMOS Decade Counter/Divider with ten decoded outputs has an internal logic arrangement of five Johnson counters with 10 decoded outputs. There are three input control signals: CLOCK, CLOCK INHIBIT, and RESET. The CLOCK signal is shaped by an internal Schmitt trigger with the output count advanced one count for each clock signal transition. This condition is true only when the CLOCK INHIBIT pin has a 0 logic. If the CLOCK INHIBIT has a 1 logic, then the external CLOCK counter advance is inhibited. The entire clock/count sequence can be restarted through a 1 logic on the RESET pin.

At the completion of each 10 count cycle there is a CARRY OUT output signal. A typical count consists of a low-to-high logic change with each output held high for the duration of one clock cycle.

Package Configuration: 16-lead DIP

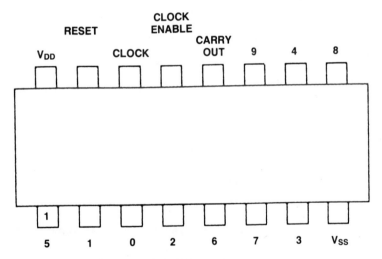

Fig. 4-1. Pin assignments for CD4017 CMOS Decade Counter/Divider.

Typical Input Capacitance: 5 pF

Clock Input Frequency: 5 - 11 MHz

Clock Pulse Width: 60 - 200 ns

Clock Inhibit-to-Clock Setup Time: 70 - 230 ns

Reset Pulse Width: 60 - 260 ns

Reset Removal Time: 150 - 400 ns

Maximum Propagation Delay Time; Decode:
650 ns at V_{DD} = 5.0 V
270 ns at V_{DD} = 10.0 V
170 ns at V_{DD} = 15.0 V

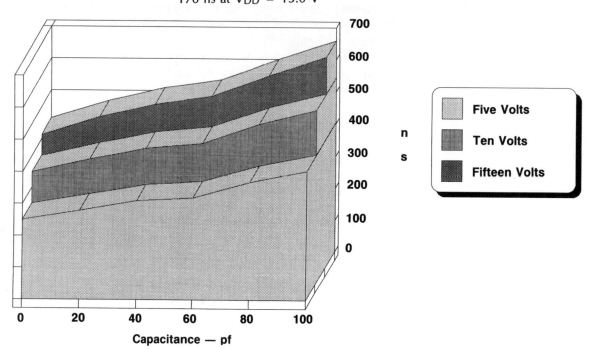

Fig. 4-2. Propagation delay for CD4017.

Maximum Propagation Delay Time; Carry Out:
600 ns at V_{DD} = 5.0 V
250 ns at V_{DD} = 10.0 V
160 ns at V_{DD} = 15.0 V

Maximum Propagation Delay Time; Reset:
530 ns at V_{DD} = 5.0 V
230 ns at V_{DD} = 10.0 V
170 ns at V_{DD} = 15.0 V

Maximum Transition Time: 200 ns at V_{DD} = 5.0 V
100 ns at V_{DD} = 10.0 V
80 ns at V_{DD} = 15.0 V

CD4018 CMOS PRESETTABLE DIVIDE-BY-'N' COUNTER

The CD4018 CMOS Presettable Divide-By-'N' Counter has an internal logic arrangement of five Johnson counters with five buffered Q outputs. There are nine input control signals: PRESET ENABLE, CLOCK, DATA, RESET, and five JAM inputs (1, 2, 3, 4, and 5). The "divide-by-N" function is determined by the routing of the buffered outputs back to the DATA pin.

A typical count consists of a low-to-high logic change with each output held high for the duration of one clock cycle. By sending a 1 logic on the RESET pin, the CD4018 can be reset to a 0 count state. Additionally, a 1 logic on the PRESET ENABLE pin causes the JAM inputs to configure the preset condition of the counter.

Package Configuration: 16-lead DIP

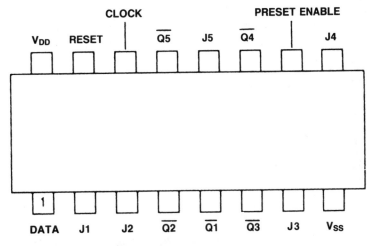

Fig. 4-3. Pin assignments for CD4018 CMOS Presettable Divide-by-'N' Counter.

Typical Input Capacitance: 7.5 pF

Clock Input Frequency: 6 - 17 MHz

Clock Pulse Width: 50 - 160 ns

Data Setup Time: 6 - 40 ns

Data Hold Time: 60 - 140 ns

Reset Pulse Width: 50 - 160 ns

Reset Removal Time: 20 - 80 ns

Maximum Propagation Delay Time; Clocked:
400 ns at V_{DD} = 5.0 V
180 ns at V_{DD} = 10.0 V
130 ns at V_{DD} = 15.0 V

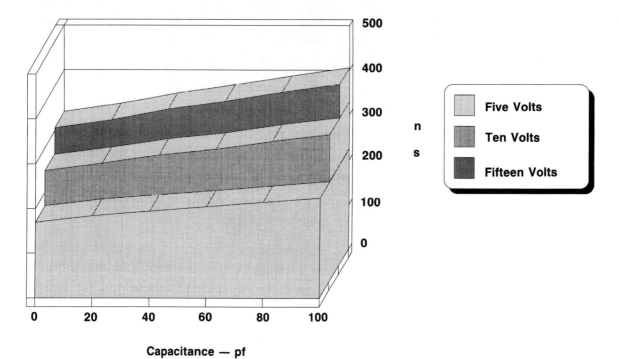

Fig. 4-4. Propagation delay for CD4018.

Maximum Propagation Delay Time; Reset:
550 ns at V_{DD} = 5.0 V
250 ns at V_{DD} = 10.0 V
180 ns at V_{DD} = 15.0 V

Maximum Transition Time:
00 ns at V_{DD} = 5.0 V
100 ns at V_{DD} = 10.0 V
80 ns at V_{DD} = 15.0 V

CD4020 CMOS 14-STAGE
RIPPLE-CARRY BINARY COUNTER/DIVIDER

The CD4020 CMOS 14-Stage Ripple-Carry Binary Counter/Divider has an internal logic arrangement of 14 master/slave flip-flops with 12 buffered outputs.

A typical count consists of a low-to-high logic change with each output held high for the duration of one clock cycle. By sending a 1 logic on the RESET pin, the CD4020 can be reset to a 0 count state.

Package Configuration: 16-lead DIP

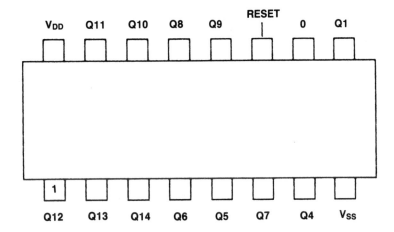

Fig. 4-5. Pin assignments for CD4020 CMOS 14-Stage Ripple-Carry Binary Counter/Divider.

Typical Input Capacitance: 7.5 pF

Input Frequency: 7 - 24 MHz

Input Pulse Width: 40 - 140 ns

Reset Pulse Width: 60 - 200 ns

Reset Removal Time: 100 - 350 ns

Maximum Propagation Delay Time; Initial Output:
360 ns at V_{DD} = 5.0 V
160 ns at V_{DD} = 10.0 V
130 ns at V_{DD} = 15.0 V

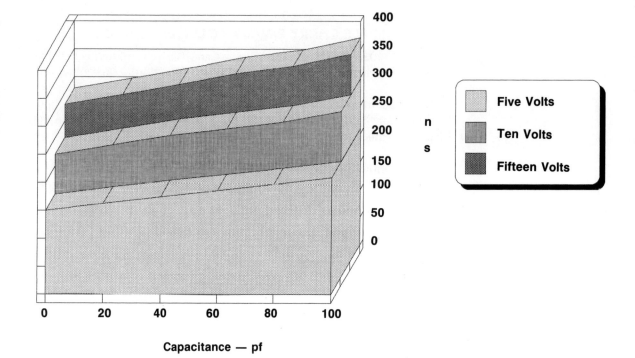

Fig. 4-6. Propagation delay for CD4020.

Maximum Propagation Delay Time; Following Outputs:
200 ns at V_{DD} = 5.0 V
80 ns at V_{DD} = 10.0 V
60 ns at V_{DD} = 15.0 V

Maximum Propagation Delay Time; Reset:
280 ns at V_{DD} = 5.0 V
120 ns at V_{DD} = 10.0 V
100 ns at V_{DD} = 15.0 V

Maximum Transition Time:
200 ns at V_{DD} = 5.0 V
100 ns at V_{DD} = 10.0 V
80 ns at V_{DD} = 15.0 V

CD4022 CMOS OCTAL COUNTER/DIVIDER

The CD4022 CMOS Octal Counter/Divider with eight decoded outputs has an internal logic arrangement of four Johnson counters with eight decoded outputs. There are three input control signals: CLOCK, CLOCK INHIBIT, and RESET. The CLOCK signal is shaped by an internal Schmitt trigger with the output count advanced one count for each clock signal transition. This condition is true only when the CLOCK INHIBIT pin has a 0 logic. If the CLOCK INHIBIT has a 1 logic, then the external CLOCK counter advance is inhibited. The entire clock/count sequence can be restarted through a 1 logic on the RESET pin.

At the completion of each eight count cycle there is a CARRY OUT output signal. A typical count consists of a low-to-high logic change with each output held high for the duration of one clock cycle.

Package Configuration: 16-lead DIP

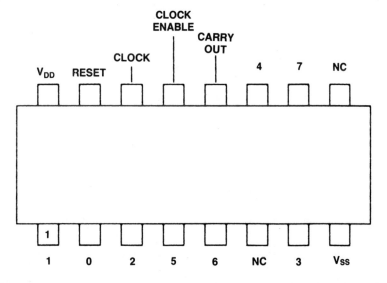

Fig. 4-7. Pin assignments for CD4022 CMOS Octal Counter/Divider with Eight Decoded Outputs.

Typical Input Capacitance: 5 pF

Clock Input Frequency: 5 - 11 MHz

Clock Pulse Width: 60 - 200 ns

Clock Inhibit-to-Clock Setup Time: 70 - 230 ns

Reset Pulse Width: 60 - 260 ns

Reset Removal Time: 150 - 400 ns

Maximum Propagation Delay Time; Decode:
650 ns at V_{DD} = 5.0 V
270 ns at V_{DD} = 10.0 V
170 ns at V_{DD} = 15.0 V

Maximum Propagation Delay Time; Carry Out:
600 ns at V_{DD} = 5.0 V
250 ns at V_{DD} = 10.0 V
160 ns at V_{DD} = 15.0 V

Maximum Propagation Delay Time; Reset:
530 ns at V_{DD} = 5.0 V
230 ns at V_{DD} = 10.0 V
170 ns at V_{DD} = 15.0 V

Maximum Transition Time:
200 ns at V_{DD} = 5.0 V
100 ns at V_{DD} = 10.0 V
80 ns at V_{DD} = 15.0 V

CD4024 CMOS SEVEN-STAGE
RIPPLE-CARRY BINARY COUNTER/DIVIDER

The CD4024 CMOS Seven-Stage Ripple-Carry Binary Counter/Divider has an internal logic arrangement of seven master/slave flip-flops with seven buffered outputs.

A typical count consists of a low-to-high logic change with each output held high for the duration of one clock cycle. By sending a 1 logic on the RESET pin, the CD4024 can be reset to a 0 count state.

Package Configuration: 14-lead DIP

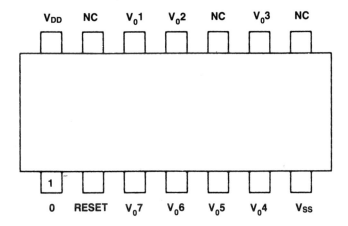

Fig. 4-8. Pin assignments for CD4024 CMOS Seven-Stage Ripple-Carry Binary Counter/Divider.

Typical Input Capacitance: 7.5 pF

Input Frequency: 7 - 24 MHz

Input Pulse Width: 40 - 140 ns

Reset Pulse Width: 60 - 200 ns

Reset Removal Time: 100 - 350 ns

Maximum Propagation Delay Time; Initial Output:
360 ns at V_{DD} = 5.0 V
160 ns at V_{DD} = 10.0 V
130 ns at V_{DD} = 15.0 V

Maximum Propagation Delay Time; Following Outputs:
200 ns at V_{DD} = 5.0 V
80 ns at V_{DD} = 10.0 V
60 ns at V_{DD} = 15.0 V

Maximum Propagation Delay Time; Reset:
280 ns at V_{DD} = 5.0 V
120 ns at V_{DD} = 10.0 V
100 ns at V_{DD} = 15.0 V

Maximum Transition Time:
200 ns at V_{DD} = 5.0 V
100 ns at V_{DD} = 10.0 V
80 ns at V_{DD} = 15.0 V

CD4026 CMOS DISPLAY
ENABLE DECADE COUNTER/DIVIDER

The CD4026 CMOS Display Enable Decade Counter/Divider is arranged with an internal logic of a five-stage Johnson decade counter and output decoder. The output from the decoder is used for driving a standard seven-segment LED display (outputs a, b, c, d, e, f, and g). Other outputs include: CARRY OUT, DISPLAY ENABLE OUT, and UNGATED "C" SEGMENT. Controlling the nature of these outputs are the CLOCK, CLOCK INHIBIT, RESET, and DISPLAY ENABLE IN.

A typical count consists of a low-to-high logic change with a 0 logic on the CLOCK INHIBIT line. Conversely, clock-induced counter advancement is inhibited with a 1 logic on the CLOCK INHIBIT line. Regardless of the current count, the display can be forced on and off through the logic of the DISPLAY ENABLE IN pin. This method is used for saving power in the circuit. The completion of a count cycle is indicated through the CARRY OUT signal. Finally, by sending a 1 logic on the RESET pin, the CD4026 can be reset to a 0 count state.

Package Configuration: 16-lead DIP

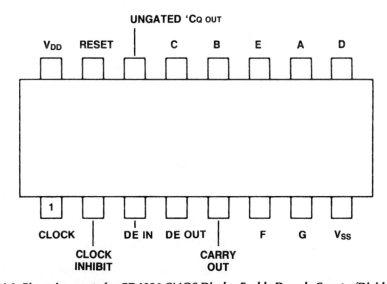

Fig. 4-9. Pin assignments for CD4026 CMOS Display Enable Decade Counter/Divider.

Maximum Input Capacitance: 7 pF

Clock Frequency: 5 - 16 MHz

Clock Pulse Width: 80 - 220 ns

Reset Pulse Width: 50 - 120 ns

Reset Removal Time: 10 - 30 ns

Maximum Propagation Delay Time; Outputs:
700 ns at V_{DD} = 5.0 V
250 ns at V_{DD} = 10.0 V
180 ns at V_{DD} = 15.0 V

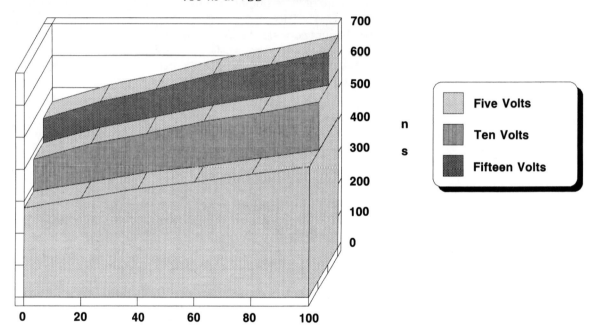

Capacitance — pf

Fig. 4-10. Propagation delay for CD4026.

Maximum Propagation Delay Time; CARRY OUT Output:
500 ns at V_{DD} = 5.0 V
200 ns at V_{DD} = 10.0 V
150 ns at V_{DD} = 15.0 V

Maximum Propagation Delay Time; Reset Outputs:
600 ns at V_{DD} = 5.0 V
250 ns at V_{DD} = 10.0 V
180 ns at V_{DD} = 15.0 V

Maximum Transition Time:
200 ns at V_{DD} = 5.0 V
100 ns at V_{DD} = 10.0 V
50 ns at V_{DD} = 15.0 V

CD4029 CMOS PRESETTABLE BINARY
OR BCD-DECADE UP/DOWN COUNTER

The CD4029 CMOS Presettable Binary or BCD-Decade Up/Down Counter is a dual mode up/down counter with four buffered outputs and one CARRY OUT output. Controlling these outputs are the inputs CLOCK, PRESET ENABLE, UP/DOWN, BINARY/DECADE, CARRY IN, and four JAM signals.

A typical count consists of a low-to-high logic change with a 0 logic on the CARRY IN and PRESET ENABLE line. Conversely, clock-induced counter advancement is inhibited with a 1 logic on either the CARRY IN or PRESET ENABLE lines. The completion of a count cycle is indicated through a 0 logic on the CARRY OUT signal. The count state of the CD4029 is set with a 1 logic on PRESET ENABLE and the correct data on the JAM inputs. Finally, by sending a 0 logic on all of the JAM pins, while the PRESET ENABLE has a 1 logic, resets the CD4029 to a 0 count state.

Mode selection is fixed through the logic of the BINARY/DECADE and the UP/DOWN pins. A 1 logic on the BINARY/DECADE pin places the CD4029 in a binary counting mode, whereas a 0 logic places the device in a BCD-decade counting mode. Likewise, the counting direction, either up or down to a fixed count, is determined through a 1 or 0 logic on the UP/DOWN pin, respectively.

Package Configuration: 16-lead DIP

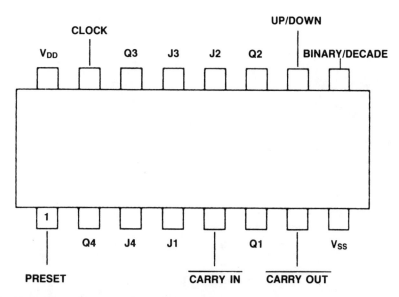

Fig. 4-11. Pin assignments for CD4029 CMOS Presettable Binary or BCD-Decade Up/Down Counter.

Maximum Input Capacitance: 7.5 pF

Clock Frequency: 4 - 11 MHz

Clock Pulse Width: 60 - 180 ns

CARRY OUT Hold Time: 25 - 50 ns

CARRY OUT Setup Time: 60 - 200 ns

Preset Pulse Width: 50 - 130 ns

Preset Removal Time: 80 - 200 ns

Maximum Propagation Delay Time; Outputs:
500 ns at V_{DD} = 5.0 V
240 ns at V_{DD} = 10.0 V
180 ns at V_{DD} = 15.0 V

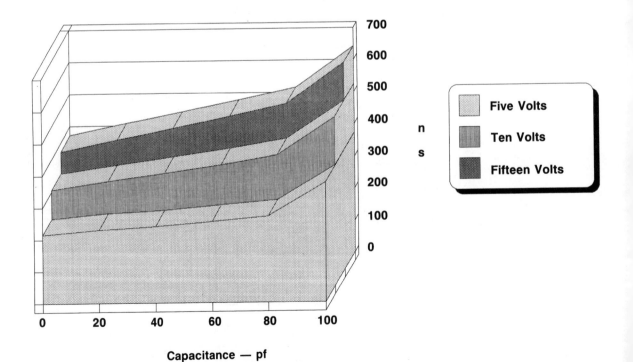

Fig. 4-12. Propagation delay for CD4029.

Maximum Propagation Delay Time; CARRY OUT Output:

560 ns at V_{DD} = 5.0 V
260 ns at V_{DD} = 10.0 V
190 ns at V_{DD} = 15.0 V

Maximum Propagation Delay Time; Reset Outputs:

470 ns at V_{DD} = 5.0 V
200 ns at V_{DD} = 10.0 V
160 ns at V_{DD} = 15.0 V

Maximum Propagation CARRY OUT Delay Time:

340 ns at V_{DD} = 5.0 V
140 ns at V_{DD} = 10.0 V
100 ns at V_{DD} = 15.0 V

Maximum Transition Time:

200 ns at V_{DD} = 5.0 V
100 ns at V_{DD} = 10.0 V
80 ns at V_{DD} = 15.0 V

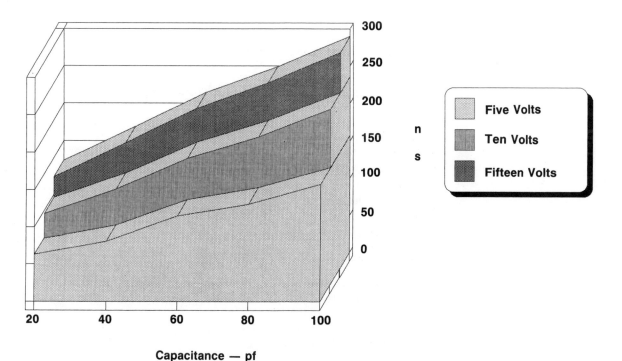

Fig. 4-13. Transition time for CD4029.

CD4040 CMOS 12-STAGE
RIPPLE-CARRY BINARY COUNTER/DIVIDER

The CD4040 CMOS 12-Stage Ripple-Carry Binary Counter/Divider has an internal logic arrangement of 12 master/slave flip-flops with 12 buffered outputs.

A typical count consists of a low-to-high logic change with each output held high for the duration of one clock cycle. By sending a 1 logic on the RESET pin, the CD4040 can be reset to a 0 count state.

Package Configuration: 16-lead DIP

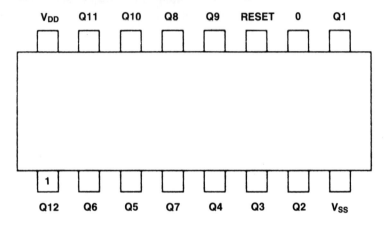

Fig. 4-14. Pin assignments for CD4040 CMOS 12-Stage Ripple-Carry Binary Counter/Divider.

Typical Input Capacitance: 7.5 pF

Input Frequency: 7 - 24 MHz

Input Pulse Width: 40 - 140 ns

Reset Pulse Width: 60 - 200 ns

Reset Removal Time: 100 - 350 ns

Maximum Propagation Delay Time; Initial Output:
360 ns at V_{DD} = 5.0 V
160 ns at V_{DD} = 10.0 V
130 ns at V_{DD} = 15.0 V

Maximum Propagation Delay Time; Following Outputs:
200 ns at V_{DD} = 5.0 V

80 ns at V_{DD} = 10.0 V
60 ns at V_{DD} = 15.0 V

Maximum Propagation Delay Time; Reset:
280 ns at V_{DD} = 5.0 V
120 ns at V_{DD} = 10.0 V
100 ns at V_{DD} = 15.0 V

Maximum Transition Time:
200 ns at V_{DD} = 5.0 V
100 ns at V_{DD} = 10.0 V
80 ns at V_{DD} = 15.0 V

CD4059 CMOS
PROGRAMMABLE DIVIDE-BY-'N' COUNTER

The CD4059 CMOS Programmable Divide-by-'N' Counter is a down counter with programmable count determined by the division of the input frequency. This ability yields a range for 'N' from 3 to 15,999. There are 16 JAM inputs that are used for selecting this count value. Based on these inputs, the output signal, OUT, is pulsed for one clock-cycle width at a rate that is set by the input frequency-'N' division.

Three inputs are used for mode selection: Ka, Kb, and Kc. There are two modes of operation that are selectable from these three inputs: the counting mode and the Master Preset mode. Mode selection is fixed through the logic of Ka, Kb, and Kc. Using all three pins for mode selection places the CD4029 in a counting mode, whereas a 0 logic on Kb and Kc places the device in Master Preset mode.

Package Configuration: 24-lead DIP

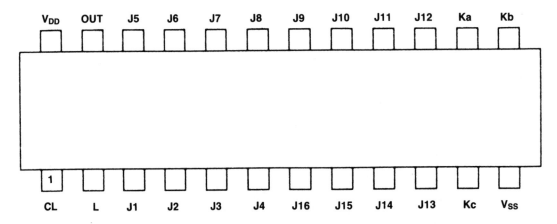

Fig. 4-15. Pin assignments for CD4059 CMOS Programmable Divide-by-'N' Counter.

Input Capacitance: 5 pF

Maximum Clock Frequency: 3 - 6 MHz

Maximum Clock Pulse Width: 100 - 200 ns

Maximum Propagation Delay Time: 360 ns at V_{DD} = 5.0 V
180 ns at V_{DD} = 10.0 V

Maximum Transition Time; High-to-Low:
 70 ns at $V_{DD} = 5.0$ V
 40 ns at $V_{DD} = 10.0$ V

Maximum Transition Time; Low-to-High:
 200 ns at $V_{DD} = 5.0$ V
 100 ns at $V_{DD} = 10.0$ V

CD4060 CMOS 14-STAGE RIPPLE-CARRY
BINARY COUNTER/DIVIDER AND OSCILLATOR

The CD4060 CMOS 14-Stage Ripple-Carry Binary Counter/Divider and Oscillator is arranged as 15 stages with buffered inputs and outputs. The first stage is an oscillator that can be used in RC circuits. The other 14 stages are separate ripple-carry binary counters with independent master/slave flip-flops. A binary count is initiated on the high-to-low pulse of the clock signal. The count in the CD4060 increases in a binary mode until the count either cycles or is terminated by the RESET input. The RESET input with a 1 logic is used for returning the device to a 0 state and inhibiting the oscillator stage.

Package Configuration: 16-lead DIP

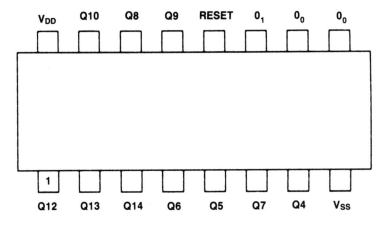

Fig. 4-16. Pin assignments for CD4060 CMOS 14-Stage Ripple-Carry Binary Counter/Divider and Oscillator.

Maximum Input Capacitance: 7.5 pF

Maximum Input Frequency: 7 - 24 MHz

Maximum Input Pulse Width: 30 - 100 ns

Maximum Reset Pulse Width: 40 - 120 ns

Maximum Oscillator Stage Frequency: 810 - 940 kHz

Maximum Propagation Delay Time; Initial:
    740 ns at V_{DD} = 5.0 V
    300 ns at V_{DD} = 10.0 V
    200 ns at V_{DD} = 15.0 V

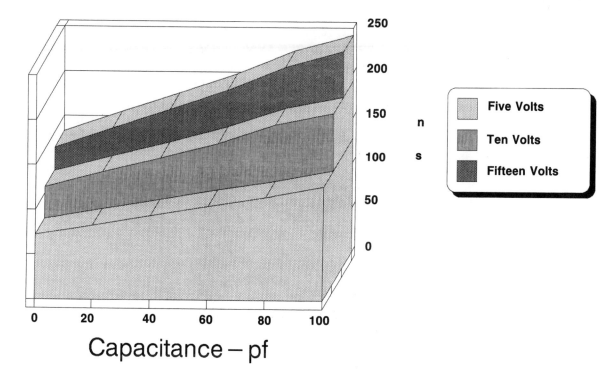

Fig. 4-17. Propagation delay for CD4060.

Maximum Propagation Delay Time; Following Outputs:

200 ns at V_{DD} = 5.0 V
100 ns at V_{DD} = 10.0 V
80 ns at V_{DD} = 15.0 V

Maximum Propagation Delay Time; Reset:

360 ns at V_{DD} = 5.0 V
160 ns at V_{DD} = 10.0 V
100 ns at V_{DD} = 15.0 V

Maximum Transition Time; High-to-Low:

200 ns at V_{DD} = 5.0 V
100 ns at V_{DD} = 10.0 V
80 ns at V_{DD} = 15.0 V

CD40103 CMOS EIGHT-STAGE 8-BIT
BINARY PRESETTABLE SYNCHRONOUS DOWN COUNTER

The CD40103 CMOS Eight-Stage 8-Bit Binary Presettable Synchronous Down Counter is arranged as a single 8-bit binary counter with one output. All of the control inputs and the single output are active with a 0 logic. The count is decremented on the low-to-high pulse of the CLOCK signal from a start of 255. This value is carried on the CARRY OUT/ZERO-DETECT output. The count can be inhibited through a 1 logic on the CARRY IN/COUNTER ENABLE pin. There are five different modes that are possible from the control inputs. Three of these modes (inhibit count, count, and preset) are synchronous modes, while the other two (preset and clear) are asynchronous functions.

Package Configuration: 16-lead DIP

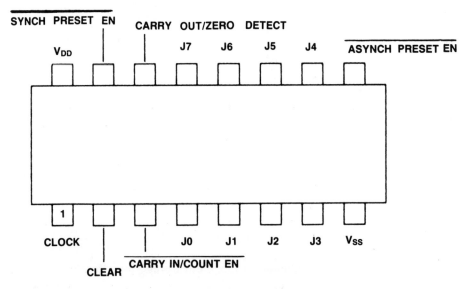

Fig. 4-18. Pin assignments for CD40103 CMOS Eight-Stage 8-Bit Binary Presettable Synchronous Down Counter.

Maximum Input Capacitance: 7.5 pF

Maximum Input Frequency: 1.4 - 4.8 MHz

Maximum Input Pulse Width: 80 - 300 ns

Maximum Propagation Delay Time; Clock-to-Output:
600 ns at V_{DD} = 5.0 V

260 ns at V_{DD} = 10.0 V
190 ns at V_{DD} = 15.0 V

Maximum Propagation Delay Time; CARRY IN-to-Output:

400 ns at V_{DD} = 5.0 V
180 ns at V_{DD} = 10.0 V
130 ns at V_{DD} = 15.0 V

Maximum Propagation Delay Time; Asynchronous Preset-to-Output:

1300 ns at V_{DD} = 5.0 V
600 ns at V_{DD} = 10.0 V
400 ns at V_{DD} = 15.0 V

Maximum Transition Time:

200 ns at V_{DD} = 5.0 V
100 ns at V_{DD} = 10.0 V
80 ns at V_{DD} = 15.0 V

CD4510 CMOS BCD
PRESETTABLE UP/DOWN COUNTER

The CD4510 CMOS BCD Presettable Up/Down Counter has an internal logic arrangement of four synchronously-clocked D-type flip-flop configured as a counter. A 1 logic on PRESET ENABLE will fix the count to any value set up on the JAM inputs. Similarly, a 1 logic on RESET will reset the counter to a zero state.

Package Configuration: 16-lead DIP

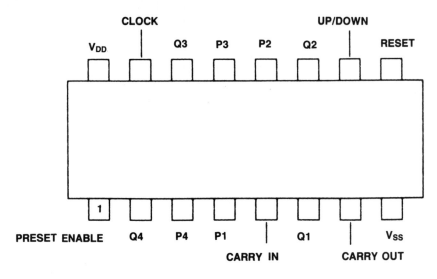

Fig. 4-19. Pin assignments for CD4510 CMOS BCD Presettable Up/Down Counter.

Maximum Input Capacitance: 7.5 pF

Maximum Input Frequency: 4 - 11 MHz

Maximum Setup Time: 5 - 12 ns

Maximum Propagation Delay Time; Clock-to-Output:
    400 ns at V_{DD} = 5.0 V
    200 ns at V_{DD} = 10.0 V
    150 ns at V_{DD} = 15.0 V

Maximum Propagation Delay Time; RESET-to-Output:
    420 ns at V_{DD} = 5.0 V
    210 ns at V_{DD} = 10.0 V
    160 ns at V_{DD} = 15.0 V

Maximum Propagation Delay Time; CLOCK-to-CARRY OUT:

480 ns at V_{DD} = 5.0 V
240 ns at V_{DD} = 10.0 V
180 ns at V_{DD} = 15.0 V

Maximum Propagation Delay Time; CARRY IN-to-CARRY OUT:

250 ns at V_{DD} = 5.0 V
120 ns at V_{DD} = 10.0 V
100 ns at V_{DD} = 15.0 V

Maximum Propagation Delay Time; RESET-to-CARRY OUT:

640 ns at V_{DD} = 5.0 V
320 ns at V_{DD} = 10.0 V
250 ns at V_{DD} = 15.0 V

Maximum Transition Time:

200 ns at V_{DD} = 5.0 V
100 ns at V_{DD} = 10.0 V
80 ns at V_{DD} = 15.0 V

CD4516 CMOS BINARY
PRESETTABLE UP/DOWN COUNTER

The CD4516 CMOS Binary Presettable Up/Down Counter has an internal logic arrangement of four synchronously-clocked D-type flip-flops configured as a counter. A 1 logic on PRESET ENABLE will fix the count to any value set up on the JAM inputs. Similarly, a 1 logic on RESET will reset the counter to a zero state.

Package Configuration: 16-lead DIP

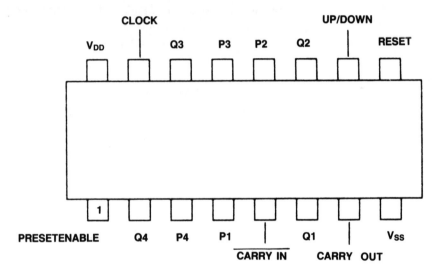

Fig. 4-20. Pin assignments for CD4516 CMOS Binary Presettable Up/Down Counter.

Maximum Input Capacitance: 7.5 pF

Maximum Input Frequency: 4 - 11 MHz

Maximum Setup Time: 5 - 12 ns

Maximum Propagation Delay Time; Clock-to-Output:
    400 ns at V_{DD} = 5.0 V
    200 ns at V_{DD} = 10.0 V
    150 ns at V_{DD} = 15.0 V

Maximum Propagation Delay Time; RESET-to-Output:
    420 ns at V_{DD} = 5.0 V
    210 ns at V_{DD} = 10.0 V
    160 ns at V_{DD} = 15.0 V

Maximum Propagation Delay Time; CLOCK-to-CARRY OUT:

480 ns at V_{DD} = 5.0 V
240 ns at V_{DD} = 10.0 V
180 ns at V_{DD} = 15.0 V

Maximum Propagation Delay Time; CARRY IN-to-CARRY OUT:

250 ns at V_{DD} = 5.0 V
120 ns at V_{DD} = 10.0 V
100 ns at V_{DD} = 15.0 V

Maximum Propagation Delay Time; RESET-to-CARRY OUT:

640 ns at V_{DD} = 5.0 V
320 ns at V_{DD} = 10.0 V
250 ns at V_{DD} = 15.0 V

Maximum Transition Time:

200 ns at V_{DD} = 5.0 V
100 ns at V_{DD} = 10.0 V
80 ns at V_{DD} = 15.0 V

CD4518 CMOS DUAL BCD UP COUNTER

The CD4518 CMOS Dual BCD Up Counter has an internal logic arrangement of two synchronous four-stage counters. Each counter stage is a D-type flip-flop with shared ENABLE and CLOCK pins for counting with either negative or positive pulse transitions. In typical applications, a positive pulse on the CLOCK line will increment the count with a 1 logic on the ENABLE input. Similarly, a 1 logic on RESET will reset the counter to a zero state.

Package Configuration: 16-lead DIP

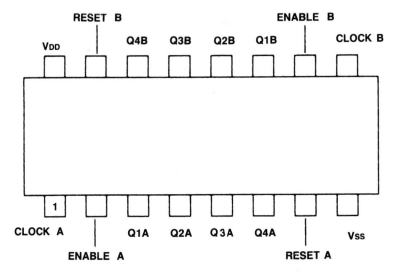

Fig. 4-21. Pin assignments for CD4518 CMOS Dual BCD Up Counter.

Maximum Input Capacitance: 7.5 pF

Maximum Clock Frequency: 3 - 8 MHz

Maximum Clock Rise or Fall Time: 5 - 15

Maximum Clock Pulse Width: 70 - 200 ns

Maximum ENABLE Pulse Width: 140 - 400 ns

Maximum RESET Pulse Width: 80 - 250 ns

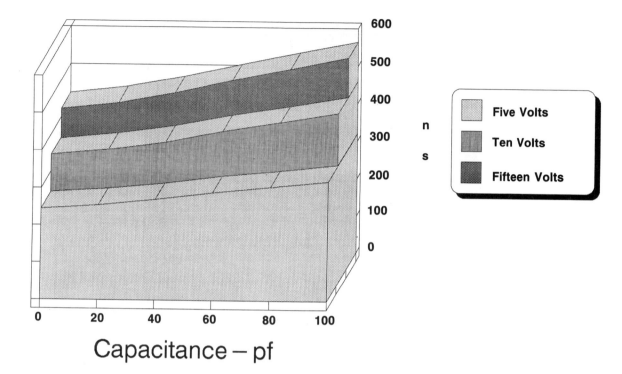

Fig. 4-22. Propagation delay for CD4518.

Maximum Propagation Delay Time; Clock-to-Output:
    560 ns at V_{DD} = 5.0 V
    230 ns at V_{DD} = 10.0 V
    160 ns at V_{DD} = 15.0 V

Maximum Propagation Delay Time; RESET-to-Output:
    650 ns at V_{DD} = 5.0 V
    225 ns at V_{DD} = 10.0 V
    170 ns at V_{DD} = 15.0 V

Maximum Transition Time:
    200 ns at V_{DD} = 5.0 V
    100 ns at V_{DD} = 10.0 V
    80 ns at V_{DD} = 15.0 V

CD4520 CMOS DUAL BINARY UP COUNTER

The CD4520 CMOS Dual Binary Up Counter has an internal logic arrangement of two synchronous four-stage counters. Each counter stage is a D-type flip-flop with shared ENABLE and CLOCK pins for counting with either negative or positive pulse transitions. In typical applications, a positive pulse on the CLOCK line will increment the count with a 1 logic on the ENABLE input. Similarly, a 1 logic on RESET will reset the counter to a zero state.

Package Configuration: 16-lead DIP

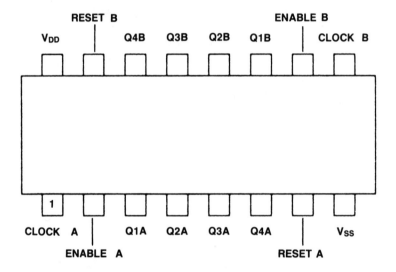

Fig. 4-23. Pin assignments for CD4520 CMOS Dual Binary Up Counter.

Maximum Input Capacitance: 7.5 pF

Maximum Clock Frequency: 3 - 8 MHz

Maximum Clock Rise or Fall Time: 5 - 15 ms

Maximum Clock Pulse Width: 70 - 200 ns

Maximum ENABLE Pulse Width: 140 - 400 ns

Maximum RESET Pulse Width: 80 - 250 ns

Maximum Propagation Delay Time; Clock-to-Output:
>   560 ns at V_{DD} = 5.0 V
>   230 ns at V_{DD} = 10.0 V
>   160 ns at V_{DD} = 15.0 V

Maximum Propagation Delay Time; RESET-to-Output:
>   650 ns at V_{DD} = 5.0 V
>   225 ns at V_{DD} = 10.0 V
>   170 ns at V_{DD} = 15.0 V

Maximum Transition Time:
>   200 ns at V_{DD} = 5.0 V
>   100 ns at V_{DD} = 10.0 V
>   80 ns at V_{DD} = 15.0 V

DIGITAL COUNTER

Second to the LED, the most visible (pun intended) GaAs (Gallium Arsenide) optoelectronic component is the common anode seven-segment LED display. This versatile 14-pin dual-inline package (DIP) is able to display all ten members of the base-10 number system based on the selective lighting of two or more of its seven segments.

In this circuit, the dual 4511 ICs (U1 and U2) receive a 4-bit code sequence from the decade counter ICs (U3 and U4). This code pattern produces a high output on a combination of the LED display's pins. The speed of displaying, advancing, and displaying the numbers is governed by the clock speed of the units decade counter IC.

The construction of the digital counter is extremely simple with the wiring schematic and parts are listed in Fig. 4-24. Only three components will need to be added to this schematic for completing the counter circuit. First, a clock circuit must be attached to the units display's decade counter IC (U3 4510). This clock circuit can be easily designed from a 4011 Quad NAND Gate.

Second, a start/stop switch, and a reset switch will be necessary for manipulating the progression of the display. These switches can be placed in two locations in the digital counter circuit: the start/stop switch between the clock and U3, and the reset switch between the power supply and the power bus of the circuit.

One final component that will enhance the performance of the digital counter is a clock rate control. This control usually is a potentiometer found in the clock circuit. By varying this control, the digital counter can be made to advance at any speed. For example, 0-99 seconds or 0-99 minutes can be selectively counted on the twin LED displays.

One final application for the digital counter is the result of combining this circuit with an optocoupler's output. This combination would create a project that is able to display the number of "trips" that have been registered between an IRED (InfraRed Light-Emitting Diode) and phototransistor. A possible arrangement for this project is given in Fig. 4-25. Based on this schematic diagram, a PCB (Printed Circuit Board) template and a parts layout are suggested in Figs. 4-26 and 4-27, respectively.

WAVE SHAPE ANALYZER

Aside from their use as logic indicators (ON or OFF), LEDs are able to duplicate the pixel representation from a low-resolution graphics display. In this application, each LED is equivalent to a single graphics screen picture element or pixel. An example of this use is presented with the wave shape analyzer, a 10 × 10 LED matrix based on a similar 5 × 7 LED matrix circuit that was described in the August 1979 issue of *Popular Electronics*.

Only three ICs are needed for generating a graphics representation of a wave shape input (see Fig. 4-28). Of these, a clock circuit sequentially pulses

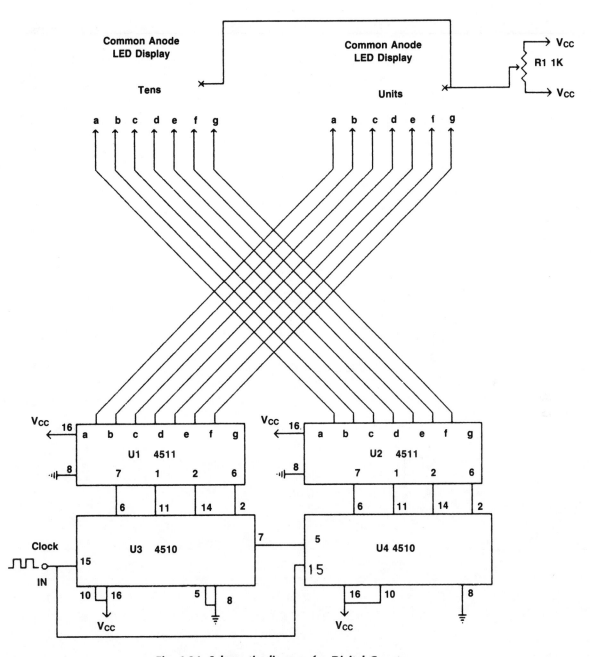

Fig. 4-24. Schematic diagram for Digital Counter.

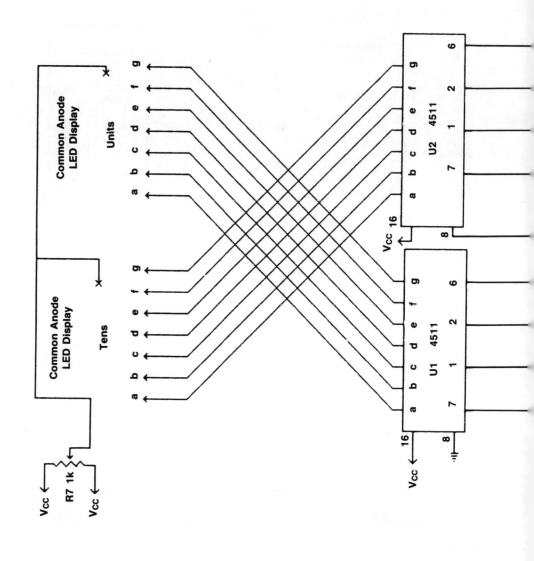

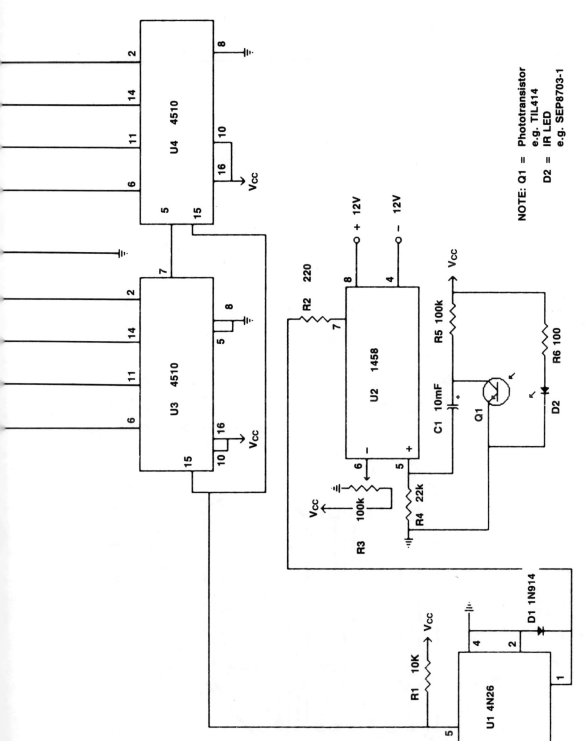

Fig. 4-25. Schematic diagram for the Digital Counter's alternate wiring scheme.

NOTE: Q1 = Phototransistor
e.g. TIL414

D2 = IR LED
e.g. SEP8703-1

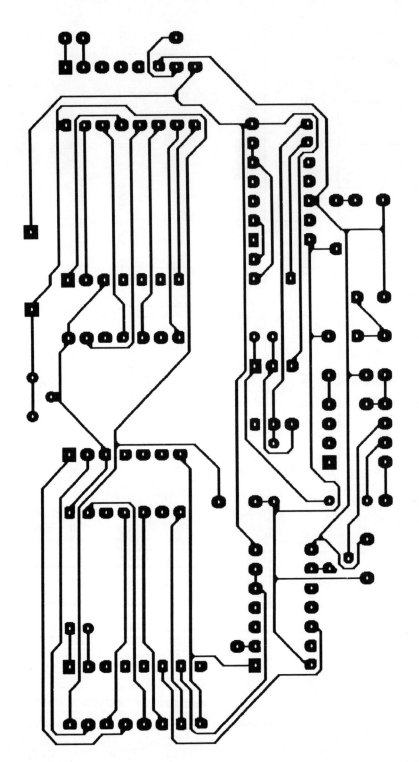

Fig. 4-26. Solder side for Digital Counter PCB template.

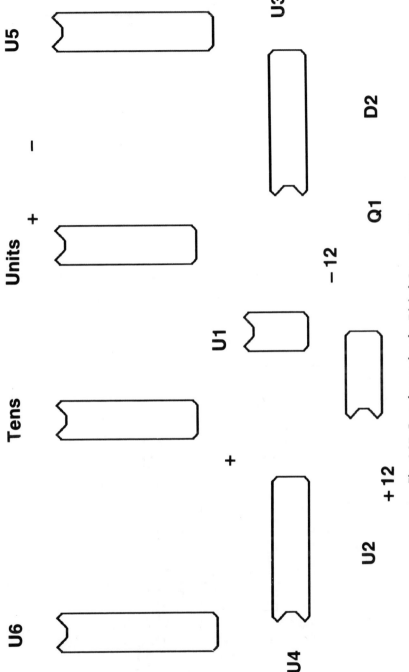

Fig. 4-27. Parts layout for the Digital Counter PCB.

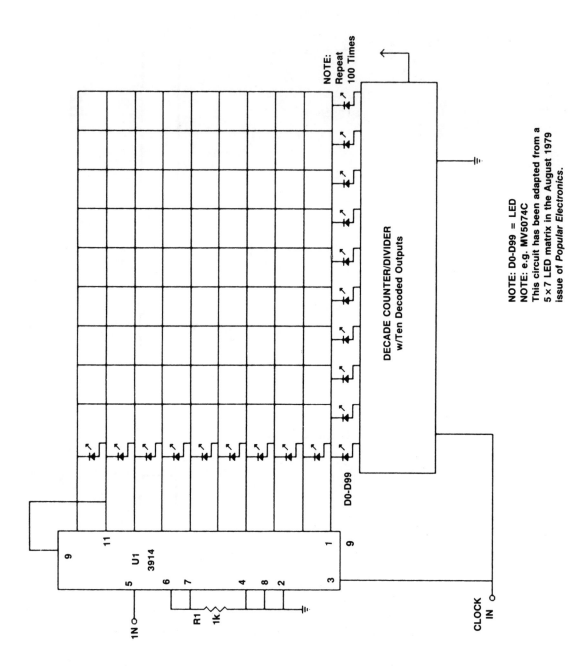

Fig. 4-28. Schematic diagram for Wave Shape Analyzer.

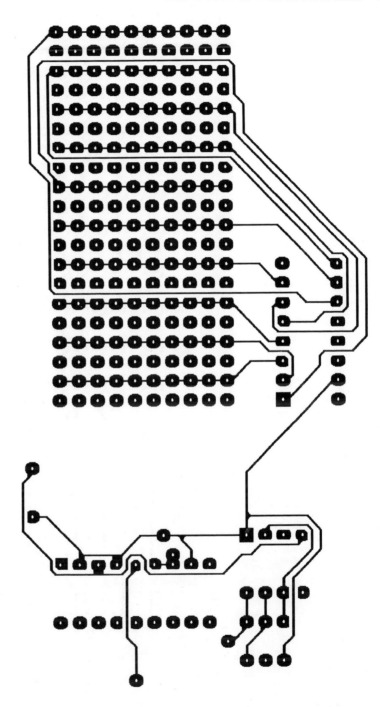

Fig. 4-29. Solder side for Wave Shape Analyzer PCB template.

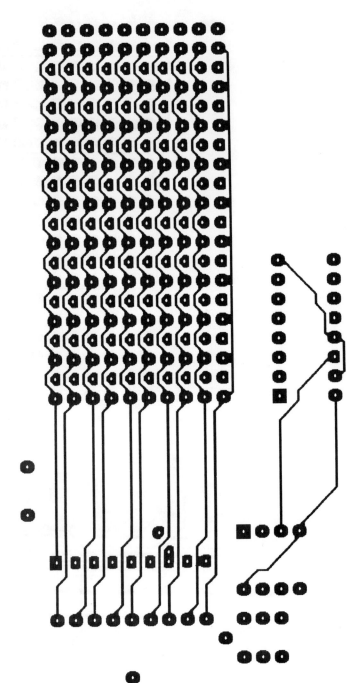

Fig. 4-30. Component side for Wave Shape Analyzer PCB template.

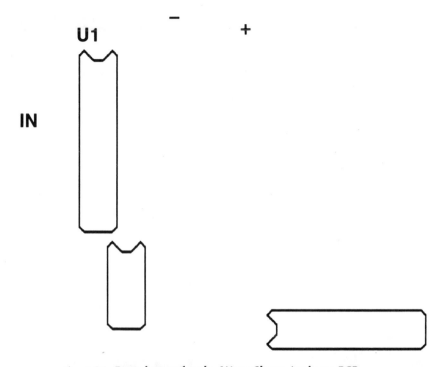

Fig. 4-31. Parts layout for the Wave Shape Analyzer PCB.

the outputs of the decade counter, while the dot driver IC (U1 3914) selectively triggers a specific row of LEDs. This triggering is based on the shape of the input waveform on pin 5 of the 3914.

The wave shape analyzer is a continuously running graphics display. A clock-based potentiometer can be used for slowing or increasing the sweep of the display. In this configuration, virtually any form of wave input can be rudimentarily visualized.

During the assembly of the wave shape analyzer be sure to observe the polarity of the LEDs. Failure to adhere to this precaution will result in an error-plagued project. Of course, soldering the 200 leads for the matrix LEDs will be an extremely taxing procedure resulting in a nightmare of twisting and overlapping cables. One solution to this spaghetti wiring is to translate the wave shape analyzer's schematic to a compact PCB design. Figs. 4-29, 4-30, and 4-31 are an example of a two- sided PCB template and a suggested parts layout, respectively. As illustrated in these templates, T-1 LEDs will fit in the provided pads. Using the larger T-1 ¾ LEDs could suffer from a congested PCB and an inability to accommodate all 100 of the LEDs.

5

Drivers

CD4056 CMOS LIQUID
CRYSTAL DISPLAY BCD-TO-SEVEN-SEGMENT
DECODER/DRIVER WITH STROBED-LATCH FUNCTION

The CD4056 CMOS Liquid Crystal Display BCD-to-Seven-Segment Decoder/Driver with Strobed-Latch Function is capable of driving a single LCD (Liquid Crystal Display) digit through seven outputs. The logic of these outputs is controlled with the DISPLAY FREQUENCY IN (DF) input. The DF input provides a 1, 0, or square-wave level-shifting output on the selected outputs. These seven outputs can be directly connected to a seven-segment LCD digit and driven by the CD4056. A CD4054 CMOS Four-Segment LCD Driver must be used with the CD4056 for providing the DF input.

In driving a single LCD digit, data are first placed on the BCD inputs with a 1 logic on the STROBE input. Subsequently, a 0 logic on the STROBE input will latch these data and select the corresponding segment output. Data latched in this manner will remain on the seven segment outputs until the STROBE input returns to a 1 logic.

Package Configuration: 16-lead DIP

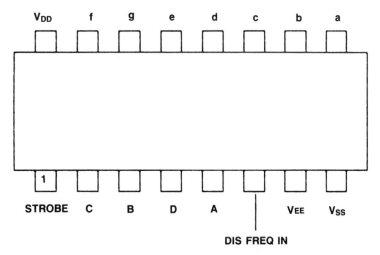

Fig. 5-1. Pin assignments for CD4056 CMOS Liquid Crystal Display BCD-to-Seven-Segment Decoder/Driver with Strobed-Latch Function.

Maximum Input Capacitance: 7.5 pF

Maximum Strobe Pulse Width: 70 - 220 ns

Maximum Data Setup Time: 70 - 220 ns

Propagation Delay Time; Input-to-Output:
1300 ns at V_{DD} = 5.0 V
1150 ns at V_{DD} = 10.0 V
750 ns at V_{DD} = 15.0 V

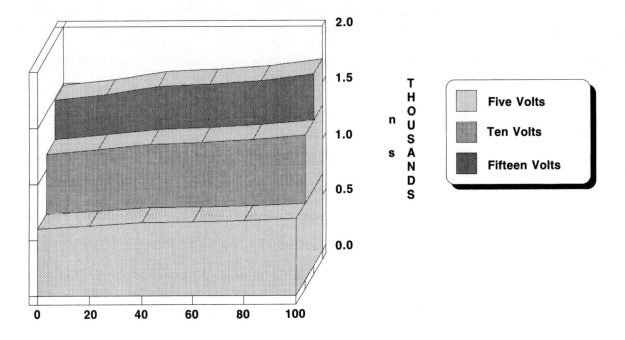

Fig. 5-2. Propagation delay for CD4056.

Maximum Output Transition Time: 200 ns at V_{DD} = 5.0 V
200 ns at V_{DD} = 10.0 V
150 ns at V_{DD} = 15.0 V

CD4511 CMOS
BCD-TO-SEVEN-SEGMENT LATCH DECODER DRIVER

The CD4511 CMOS BCD-to-Seven-Segment Latch Decoder Driver is capable of either directly driving or multiplexing a single LED (Light-Emitting Diode) display through seven NPN bipolar transistor outputs. This LED display must be of the common- cathode configuration.

There are three control inputs that are 0 logic active: LT (Lamp Test), BL (BLanking), and LE/STROBE (Latch Enable). Each of these inputs controls the display test, the display intensity, and the storing or strobing of BCD data, respectively.

Package Configuration: 16-lead DIP

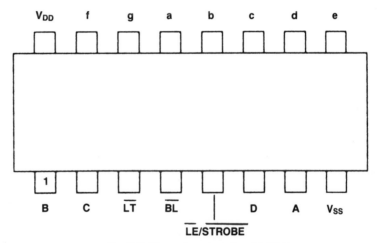

Fig. 5-3. Pin assignments for CD4511
CMOS BCD-to-Seven-Segment Latch Decoder Driver.

Maximum Input Capacitance: 7.5 pF

Maximum Strobe Pulse Width: 50 - 200 ns

Maximum Data Setup Time: 20 - 150 ns

Propagation Data Delay Time; High-to-Low:
  1040 ns at V_{DD} = 5.0 V
  420 ns at V_{DD} = 10.0 V
  300 ns at V_{DD} = 15.0 V

Propagation Blanking Delay Time; High-to-Low:
700 ns at V_{DD} = 5.0 V
350 ns at V_{DD} = 10.0 V
250 ns at V_{DD} = 15.0 V

Propagation Lamp Test Delay Time; High-to-Low:
500 ns at V_{DD} = 5.0 V
250 ns at V_{DD} = 10.0 V
170 ns at V_{DD} = 15.0 V

Maximum Transition Time; High-to-Low:
80 ns at V_{DD} = 5.0 V
60 ns at V_{DD} = 10.0 V
50 ns at V_{DD} = 15.0 V

CD4543 CMOS BCD-TO-SEVEN-SEGMENT
LATCH/DECODER/DRIVER FOR LIQUID CRYSTAL DISPLAYS

The CD4543 CMOS BCD-to-Seven-Segment Latch/Decoder/Driver for Liquid Crystal Displays is capable of directly driving a single LCD digit, although LED displays can also be interfaced to its seven outputs. The functions of the CD4543 are similar to those of the CD4056 with the exception of the CD4543 device's omission of the level-shifting function (a display blanking function has been substituted for this missing function).

When used with LED displays the CD4543 must have either a high logic on its PHASE input for common-cathode displays or a low logic for common-anode types. Similarly, the PHASE input is used for receiving a square wave input during LCD application.

Package Configuration: 16-lead DIP

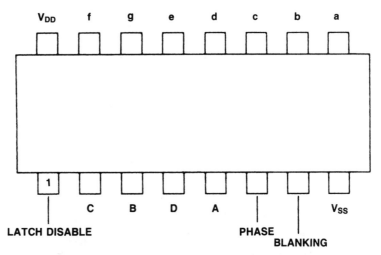

*Fig. 5-4. Pin assignments for CD4543 CMOS
BCD-to-Seven-Segment Latch/Decoder/Driver for Liquid Crystal Displays.*

Maximum Input Capacitance: 7.5 pF

Maximum LATCH DISABLE Pulse Width: 40 - 250 ns

Maximum Address Setup Time: 10 - 60 ns

Propagation Delay Time; High-to-Low: 1200 ns at V_{DD} = 5.0 V
400 ns at V_{DD} = 10.0 V
300 ns at V_{DD} = 15.0 V

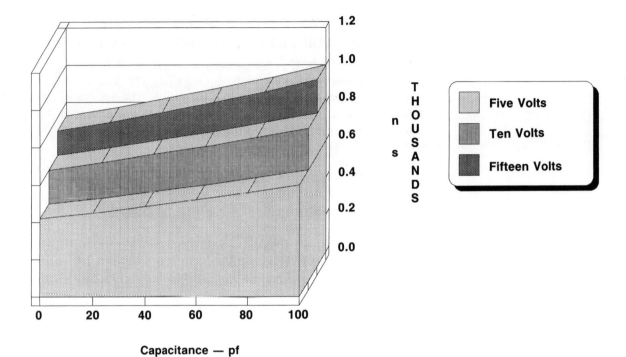

Capacitance — pf

Fig. 5-5. Propagation delay for CD4543.

Propagation Delay Time; Low-to-High: 1000 ns at V_{DD} = 5.0 V
                                              400 ns at V_{DD} = 10.0 V
                                              300 ns at V_{DD} = 15.0 V

Maximum Transition Time: 360 ns at V_{DD} = 5.0 V
                             180 ns at V_{DD} = 10.0 V
                             130 ns at V_{DD} = 15.0 V

INTERSIL ICL7107 CMOS 3½-DIGIT
LED SINGLE-CHIP A/D CONVERTER

The Intersil ICL7107 CMOS 3½-Digit LED Single-Chip A/D Converter is a complete self-contained seven-segment decoder, driver, reference, and clock capable of directly driving a single 3½-digit LED display. No other support devices are needed for interfacing the ICL7107 directly to an LED display.

A companion device to the ICL7107 is the ICL7106 that shares virtually all of the same characteristics except the ICL7106 will directly drive a 3½-digit LCD. Additionally, the ICL7106 is operational from a single power supply, while the ICL7107 requires two power supplies for properly driving the LED display.

Package Configuration: 40-lead DIP

Common Mode Rejection Ratio: CMRR = 50 mV/V

Standard System Noise: 15 mV

Maximum Input Leakage Current: 10 pA

Maximum V+ Power Supply Current: 1.8 mA

Maximum V− Power Supply Current: 1.8 mA

INTERSIL ICM7226A CMOS EIGHT-DIGIT
LED MULTI-FUNCTION FREQUENCY COUNTER/TIMER

The Intersil ICM7226A CMOS Eight-Digit LED Multi-Function Frequency Counter/Timer is a complete, self-contained, universal counter with an LED display driver. The internal structure of the ICM7226A features a seven-segment decoder, driver, digit multiplexer, high-frequency oscillator, decade timebase counter, and an eight decade data counter. Only common-anode LED displays with a peak segment current of 25 mA can be connected to the ICM7226A, while common-cathode LED displays (with a peak segment current of 12 mA) can be used with the ICM7226B.

In either case, the LED displays are multiplexed at 500 Hz and a duty cycle of 12.2 percent. Other features of the display driver section of the ICM7226x includes a blanking of leading zeros, kHz units for frequency displays, and mS units used for time measurements.

Package Configuration: 40-lead DIP

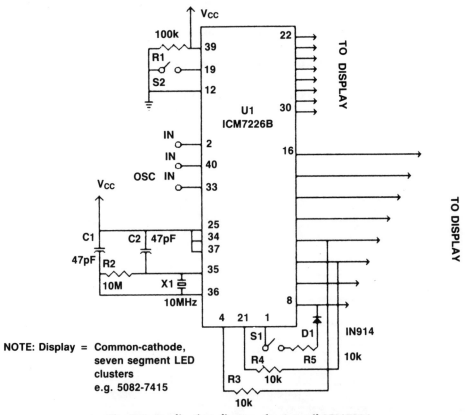

NOTE: Display = Common-cathode, seven segment LED clusters e.g. 5082-7415

Fig. 5-6. Application diagram for Intersil ICM7226
CMOS Eight-Digit LED Multi-Function Frequency Counter/Timer.

Time Between Samples: 200 ms

Input Charge Rate: 15 mV/ms

Digit Driver Output Current: −0.3 - 180 mA

Maximum Input Low Voltage: 0.8 V

INTERSIL ICM7243 CMOS EIGHT-CHARACTER LED MICROPROCESSOR-COMPATIBLE DISPLAY DRIVER

The Intersil ICM7243 CMOS Eight-Character LED Microprocessor-Compatible Display Driver is an LED display driver interface between 14- or 16-segment LED displays and a microprocessor, or a digital system. The internal structure of the ICM7243 features a 64-character ASCII (American Standard Code for Information Interchange) driver, digit multiplexer, and 8 × 6 memory. Only common-cathode 16-segment LED displays can be connected to the ICM7243A, while common-cathode 14-segment LED displays can be used with the ICM7243B.

During operation, the internal memory of the ICM7243x is loaded with the six-bit ASCII code. The subsequent placement of this display data is determined by two modes of operation. In the SERIAL ACCESS mode of operation, the first character is displayed in the furthest left character position. The following characters are then placed to the right of this starting digit.

In the RANDOM ACCESS mode of operation, the controlling microprocessor is used for selecting the exact location for each character entry. The selection of either of these modes of operation is dictated through the logic on the MODE line of the ICM7243x.

Package Configuration: 40-lead DIP

Input Capacitance: 5 pF

Data Setup Time: 150 ns

Address Setup Time: 15 ns

Display Scan Rate: 400 Hz

Character Drive Current: 190 mA

Segment Drive Current: 19 mA

Maximum Transition Pulse Time: 100 ns

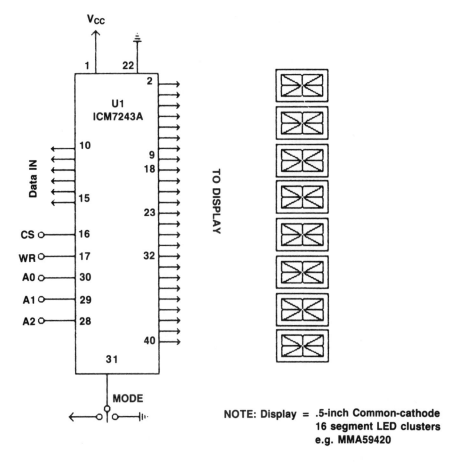

Fig. 5-7. Application diagram for Intersil ICM7243 CMOS Eight-Character LED Microprocessor-Compatible Display Driver.

A/D CONVERTER

Intersil manufactures a series of display ICs, each of which are capable of driving a segmented LED display. The first of these display ICs is the ICL7107. This is an A/D (Analog/Digital) converter chip that can directly drive a 3½-digit 7-segment LED display.

The A/D Converter uses a handful of support components for converting a voltage input into a digital display (see Fig. 5-8). All of the other active display logic circuitry, decoders, drivers, and clock, are self-contained within the ICL7107. In addition to the support components, three common anode 7-segment displays, along with a common anode overflow display are used for displaying the voltage input.

During operation of the A/D Converter, two separate supply voltages are required by the ICL7107 (U1). These + 5.0 V and − 5.0 V are applied to pins 1 and 26, respectively. A voltage variation to a maximum of + 6.0 V and − 9.0 V is tolerable.

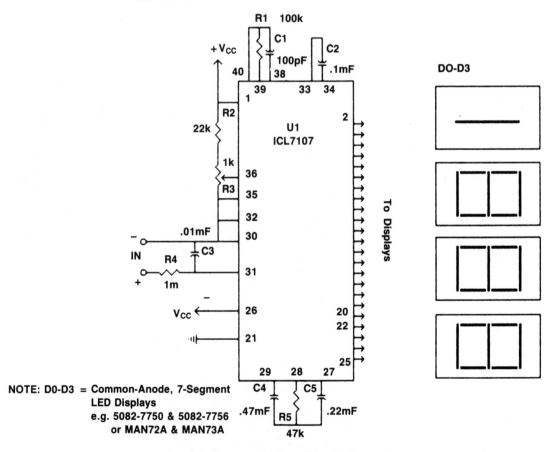

Fig. 5-8. Schematic diagram for A/D Converter.

6

Multiplexers/Demultiplexers

CD4016 CMOS QUAD BILATERAL SWITCH

The CD4016 CMOS Quad Bilateral Switch has an internal logic of four separate bilateral switches, each of which is controlled by a separate control input (CONTROL x). Although not a dedicated analog multiplexer device, the CD4016 can be easily adapted to simplified digital and analog multiplexing duties.

A pin-for-pin replacement for the CD4016 is the CD4066 Quad Bilateral Switch. Two important design differences separate these two devices. First, the CD4066 has a lower on-state resistance. Second, in the CD4066, this on-state resistance is uniform throughout the input signal range.

Package Configuration: 14-lead DIP

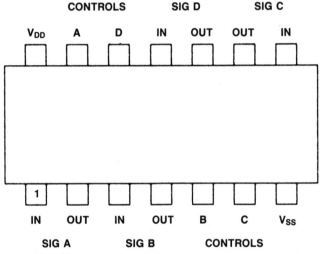

Fig. 6-1. Pin assignments for CD4016 CMOS Quad Bilateral Switch.

Crosstalk Frequency: 0.9 MHz

Control Repetition Rate: 10 MHz

Total Harmonic Distortion: 0.4 %

Maximum Quiescent Device Current: .25 mA at V_{DD} = 5.0 V
.5 mA at V_{DD} = 10.0 V
1 mA at V_{DD} = 15.0 V

Propagation Delay Time: 100 ns at V_{DD} = 5.0 V
40 ns at V_{DD} = 10.0 V
30 ns at V_{DD} = 15.0V

CD4051 CMOS SINGLE EIGHT-CHANNEL ANALOG MULTIPLEXER/DEMULTIPLEXER

The CD4051 CMOS Single Eight-Channel Analog Multiplexer/Demultiplexer has an internal structure consisting of eight analog switches that are controlled via three digital inputs (A, B, and C), and an inhibit input (INH). The logic on the three control inputs determines which of the eight channels will be activated, and interfaces one of the eight inputs (CHANNELS IN/OUT) to the output (COM OUT/IN).

Package Configuration: 16-lead DIP

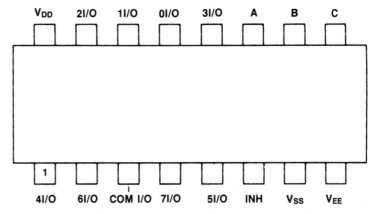

Fig. 6-2. Pin assignments for CD4051 CMOS Single Eight-Channel Analog Multiplexer/Demultiplexer.

Output Capacitance: 30 pF

Total Harmonic Distortion: 0.12 - 0.3 %

Propagation Signal Delay Time: 20 - 60 ns

Cutoff Frequency; Channel On: 30 MHz

Feedthrough Frequency; Channel Off:8 MHz

Maximum Control Input Low Voltage:
    1.5 V at V_{DD} = 5.0 V
    3.0 V at V_{DD} = 10.0 V
    4.0 V at V_{DD} = 15.0 V

Maximum Quiescent Signal Device Current:
    5 mA at V_{DD} = 5.0 V
    10 mA at V_{DD} = 10.0 V
    20 mA at V_{DD} = 15.0 V

CD4052 CMOS DIFFERENTIAL
FOUR-CHANNEL ANALOG MULTIPLEXER/DEMULTIPLEXER

The CD4052 CMOS Differential Four-Channel Analog Multiplexer/Demultiplexer has an internal structure consisting of eight analog switches that are controlled via two digital inputs (A and B) and an inhibit input (INH). The logic on the two control inputs determines which of four pairs of channels will be activated, and interfaces one of the inputs (X or Y CHANNELS IN/OUT) to the output (COMMON "X" or "Y" OUT/IN).

Package Configuration: 16-lead DIP

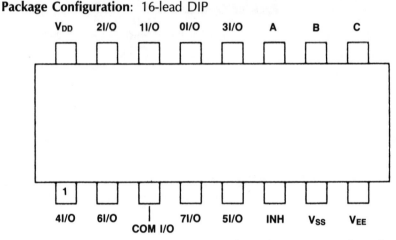

Fig. 6-3. Pin assignments for CD4052 CMOS Differential Four-Channel Analog Multiplexer/Demultiplexer.

Output Capacitance: 18 pF

Total Harmonic Distortion: 0.12 - 0.3 %

Propagation Signal Delay Time: 20 - 60 ns

Cutoff Frequency; Channel On: 25 MHz

Feedthrough Frequency; Channel Off: 10 MHz

Maximum Control Input Low Voltage:
1.5 V at V_{DD} = 5.0 V
3.0 V at V_{DD} = 10.0 V
4.0 V at V_{DD} = 15.0 V

Maximum Quiescent Signal Device Current:
5 mA at V_{DD} = 5.0 V
10 mA at V_{DD} = 10.0 V
20 mA at V_{DD} = 15.0 V

CD4053 CMOS TRIPLE TWO-CHANNEL ANALOG MULTIPLEXER/DEMULTIPLEXER

The CD4053 CMOS Triple Two-Channel Analog Multiplexer/Demultiplexer has an internal structure consisting of six SPDT (Single-Pole, Double-Throw) analog switches controlled via three digital inputs (A, B, and C), and an inhibit input (INH). These three control inputs determine which of three pairs of channels will be activated.

Package Configuration: l6-lead DIP

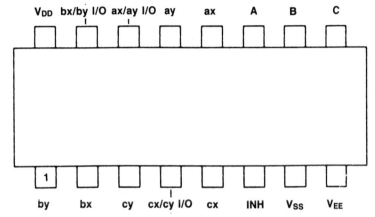

Fig. 6-4. Pin assignments for CD4053 CMOS Triple Two-Channel Analog Multiplexer/Demultiplexer.

Output Capacitance: 9 pF

Total Harmonic Distortion: 0.12 - 0.3 %

Propagation Signal Delay Time: 20 - 60 ns

Cutoff Frequency; Channel On: 20 MHz

Feedthrough Frequency; Channel Off: 12 MHz

Maximum Control Input Low Voltage:
    1.5 V at V_{DD} = 5.0 V
    3.0 V at V_{DD} = 10.0 V
    4.0 V at V_{DD} = 15.0 V

Maximum Quiescent Signal Device Current:
    5 mA at V_{DD} = 5.0 V
    10 mA at V_{DD} = 10.0 V
    20 mA at V_{DD} = 15.0 V

CD4067 CMOS SINGLE 16-CHANNEL ANALOG MULTIPLEXER/DEMULTIPLEXER

The CD4067 CMOS Single 16-Channel Analog Multiplexer/Demultiplexer consists of 16 analog switches controlled via three digital inputs (A, B, and C), and an inhibit input (INH). The logic on three control inputs determines which of the switches will be activated. A high logic on the INH input forces all switches off.

Package Configuration: 24-lead DIP

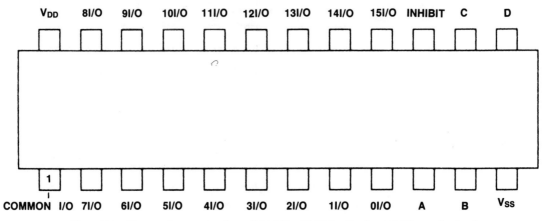

Fig. 6-5. Pin assignments for CD4067 CMOS Single 16-Channel Analog Multiplexer/Demultiplexer.

Output Capacitance: 55 pF

Total Harmonic Distortion: 0.12 - 0.3 %

Propagation Signal Delay Time: 20 - 60 ns

Cutoff Frequency; Channel On: 14 MHz

Feedthrough Frequency; Channel Off: 20 MHz

Maximum Control Input Low Voltage:
　　1.5 V at V_{DD} = 5.0 V
　　3.0 V at V_{DD} = 10.0 V
　　4.0 V at V_{DD} = 15.0 V

Maximum Quiescent Signal Device Current:
　　5 mA at V_{DD} = 5.0 V
　　10 mA at V_{DD} = 10.0 V
　　20 mA at V_{DD} = 15.0 V

INTERSIL IH6108 CMOS
EIGHT-CHANNEL ANALOG MULTIPLEXER

The Intersil IH6108 CMOS Eight-Channel Analog Multiplexer contains eight analog switches. Three address inputs and one enable input control the nature of the eight outputs. The enable input is actually a switch that is used for enabling or disabling all of the outputs. Therefore, any of the outputs can be selected by the address inputs when the enable output has a 1 logic.

Package Configuration: 16-lead DIP

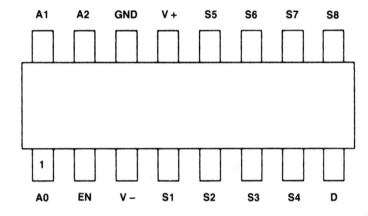

Fig. 6-6. Pin assignments for IH6108 CMOS Eight-Channel Analog Multiplexer.

Off Isolation: 60 dB

V+ Power Supply Current: 40 mA

V– Power Supply Current: 2 mA

Standby Current: 1 mA

Transition Time: 0.3 ms

D/A CONVERTER

With a modest number of support components, a variable D/A converter can be designed from the CD4016 Quad Bilateral Switch device (see Fig. 6-7). This variable output can be further interfaced with numerous voltage-sensitive devices for generating an artificial random waveform.

Unlike the A/D Converter demonstrated in the previous chapter, the D/A Converter takes a 4-bit input, and outputs an analog voltage signal. The supply for this digital input can be either hardwired for a specific analog voltage, or produced in a programmed fashion. Virtually any up-counter device can be used for providing the BCD count required in this application.

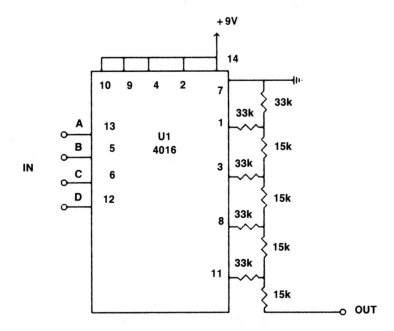

Fig. 6-7. Schematic diagram for D/A Converter.

7

Multivibrators, Schmitt Triggers, and Flip-Flops

CD4047 CMOS LOW-POWER
MONOSTABLE/ASTABLE MULTIVIBRATOR

The CD4047 CMOS Low-Power Monostable/Astable Multivibrator is a gatable astable multivibrator with a positive- or negative-edge triggered monostable multivibrator capability. There are five buffered inputs, ASTABLE, −TRIGGER, +TRIGGER, EXTERNAL RESET, and RETRIGGER. While all of these inputs are active high logic inputs, there is another ASTABLE input that is an active low logic input. These six inputs control three buffered outputs: Q, Q (active low logic), and OSC OUT. Both a timing capacitor input, and a timing resistor input are required in either an astable or a monostable application.

Package Configuration: 14-lead DIP

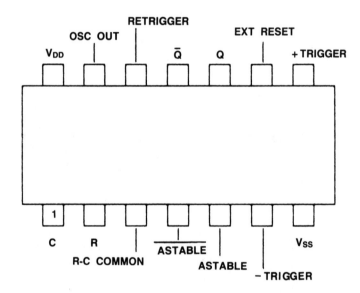

*Fig. 7-1. Pin assignments for CD4047
CMOS Low-Power Monostable/Astable Multivibrator.*

Maximum Input Capacitance: 7.7 pF

Maximum Input Pulse Width; +TRIGGER, −TRIGGER:
    400 ns at V_{DD} = 5.0 V
    160 ns at V_{DD} = 10.0 V
    100 ns at V_{DD} = 15.0 V

Maximum Input Pulse Width; RETRIGGER:

600 ns at V_{DD} = 5.0 V
230 ns at V_{DD} = 10.0 V
150 ns at V_{DD} = 15.0 V

Maximum Propagation Delay Time; +TRIGGER-to-outputs:

1000 ns at V_{DD} = 5.0 V
450 ns at V_{DD} = 10.0 V
300 ns at V_{DD} = 15.0 V

CD4098 CMOS DUAL MONOSTABLE MULTIVIBRATOR

The CD4098 CMOS Dual Monostable Multivibrator contains two fixed one-shot retriggerable, resettable monostable multivibrators. A separate set of control inputs, and buffered outputs are provided for each monostable multivibrator. Both positive- and negative-edge trigger inputs, $+TR$ and $-TR$, respectively, are available. One final duplicated control input is the active high logic RESET. Both output pulse termination, and power on output pulse mute functions are controlled via the RESET pin. The output from these inputs is two buffered outputs Q and $\overline{Q}$ (active low logic). Two final inputs control the output pulse widths with a timing capacitor input and a timing resistor input. These timing inputs are duplicated in both monostable multivibrators.

Package Configuration: 16-lead DIP

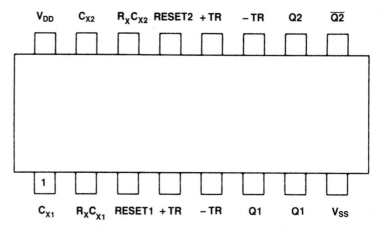

Fig. 7-2. Pin assignments for CD4098 CMOS Dual Monostable Multivibrator.

Maximum Input Capacitance: 7.5 pF

Maximum Reset Pulse Width; Rx = 100k, Cx = 15 pF:
    200 ns at V_{DD} = 5.0 V
    80 ns at V_{DD} = 10.0 V
    60 ns at V_{DD} = 15.0 V

Maximum Trigger Pulse Width:

    140 ns at V_{DD} = 5.0 V
    60 ns at V_{DD} = 10.0 V
    40 ns at V_{DD} = 15.0 V

Maximum Propagation Delay Time; +TR-to-outputs:

   500 ns at V_{DD} = 5.0 V
   250 ns at V_{DD} = 10.0 V
   200 ns at V_{DD} = 15.0 V

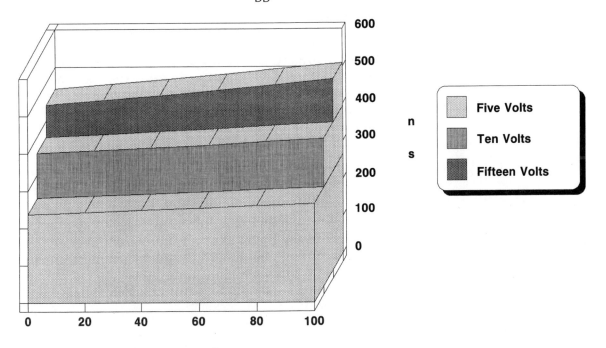

Fig. 7-3. Propagation delay for CD4098.

Maximum Propagation Delay Time;   Reset:

   450 ns at V_{DD} = 5.0 V
   250 ns at V_{DD} = 10.0 V
   150 ns at V_{DD} = 15.0 V

Maximum Transition Time; Cx = 15 pF, High-to-Low:

   200 ns at V_{DD} = 5.0 V
   100 ns at V_{DD} = 10.0 V
   80 ns at V_{DD} = 15.0 V

Maximum Transition Time; Cx = .01 mf, High-to-Low:

   300 ns at V_{DD} = 5.0 V
   150 ns at V_{DD} = 10.0 V
   130 ns at V_{DD} = 15.0 V

CD4538 CMOS DUAL
PRECISION MONOSTABLE MULTIVIBRATOR

The CD4538 CMOS Dual Precision Monostable Multivibrator contains two stable one-shot retriggerable, resettable, fixed- voltage timing monostable multivibrators. A separate set of control inputs and outputs are provided for each monostable multivibrator. Both positive- and negative-edge trigger inputs, +TR and −TR, respectively, are available. One final duplicated control input is the active high logic RESET. Both output pulse termination, and power on output pulse mute functions are controlled via the RESET pin. The output from these inputs is two buffered outputs Q and $\overline{Q}$ (active low logic). Two final inputs control the output pulse widths through a timing capacitor input, and a timing resistor input. These timing inputs are duplicated in both monostable multivibrators.

Package Configuration: 16-lead DIP

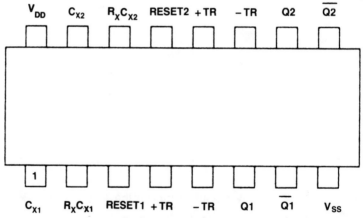

Fig. 7-4. Pin assignments for CD4538 CMOS Dual Precision Monostable Multivibrator.

Maximum Input Capacitance: 7.5 pF

Maximum Reset Pulse Width:
140 ns at V_{DD} = 5.0 V
80 ns at V_{DD} = 10.0 V
60 ns at V_{DD} = 15.0 V

Maximum Output Pulse Width; Rx = 100k Cx = .1 mF:
10.5 ms at V_{DD} = 5.0 V
10.6 ms at V_{DD} = 10.0 V
10.6 ms at V_{DD} = 15.0 V

Maximum Output Pulse Width; Rx = 100k Cx = 10 mF:

1.06 s at V_{DD} = 5.0 V
1.06 s at V_{DD} = 10.0 V
1.07 s at V_{DD} = 15.0 V

Maximum Propagation Delay Time; +TR-to-outputs:

600 ns at V_{DD} = 5.0 V
300 ns at V_{DD} = 10.0 V
220 ns at V_{DD} = 15.0 V

Maximum Propagation Delay Time; Reset:

500 ns at V_{DD} = 5.0 V
250 ns at V_{DD} = 10.0 V
190 ns at V_{DD} = 15.0 V

Maximum Transition Time; High-to-Low:

200 ns at V_{DD} = 5.0 V
100 ns at V_{DD} = 10.0 V
80 ns at V_{DD} = 15.0 V

CD4093 CMOS QUAD
TWO-INPUT NAND SCHMITT TRIGGER

The CD4093 CMOS Quad Two-Input NAND Schmitt Trigger has an internal logic of four, two-input NAND gates with Schmitt trigger circuits on both inputs. Each gate exhibits a distinct switching point for both positive-, as well as negative-transition signals. A hysteresis voltage is represented as the difference between these positive and negative voltages.

Package Configuration: 14-lead DIP

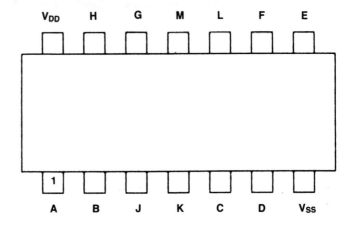

Fig. 7-5. Pin assignments for CD4093 CMOS Quad Two-Input NAND Schmitt Trigger.

Maximum Input Capacitance: 7.5 pF

Maximum Hysteresis Voltage; Both Inputs, A and B:
  1.6 V at V_{DD} = 5.0 V
  3.4 V at V_{DD} = 10.0 V
  5.0 V at V_{DD} = 15.0 V

Maximum Positive Trigger Threshold Voltage; A Input:
  3.6 V at V_{DD} = 5.0 V
  7.1 V at V_{DD} = 10.0 V
10.8 V at V_{DD} = 15.0 V

Maximum Positive Trigger Threshold Voltage; B Input:
  4.0 V at V_{DD} = 5.0 V
  8.2 V at V_{DD} = 10.0 V
12.7 V at V_{DD} = 15.0 V

Maximum Negative Trigger Threshold Voltage; A Input:
2.8 V at V_{DD} = 5.0 V
5.2 V at V_{DD} = 10.0 V
7.4 V at V_{DD} = 15.0 V

Maximum Negative Trigger Threshold Voltage; B Input:
3.2 V at V_{DD} = 5.0 V
6.6 V at V_{DD} = 10.0 V
9.6 V at V_{DD} = 15.0 V

Maximum Propagation Delay Time; High-to-Low:
380 ns at V_{DD} = 5.0 V
180 ns at V_{DD} = 10.0 V
130 ns at V_{DD} = 15.0 V

CD4013 CMOS DUAL D-TYPE FLIP-FLOP

The CD4013 CMOS Dual D -Type Flip-Flop has an internal logic of four data-type flip-flops. These flip-flops are completely static with their states remaining intact with a 1 or 0 logic clock. Each flip-flop possesses separate set, data, clock, and reset inputs along with dual outputs (one output is inverted). All inputs are high logic active with the set and reset inputs being independent of the clock.

Package Configuration: 14-lead DIP

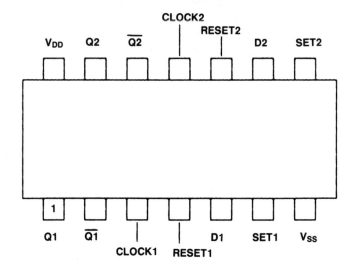

Fig. 7-6. Pin assignments for CD4013 CMOS Dual D -Type Flip-Flop.

Maximum Input Capacitance: 7.5 pF

Clock Input Frequency: 3.5 - 24 MHz

Clock Pulse Width: 20 - 140 ns

Set Pulse Width: 25 - 180 ns

Reset Pulse Width: 25 - 180 ns

Data Setup Time: 7 - 40 ns

Maximum Clock Input Rise & Fall Time: 2 - 70 ms

Maximum Propagation Delay Time; Clock-, Set-, or Reset-to-Ouputs:
380 ns at V_{DD} = 5.0 V
130 ns at V_{DD} = 10.0 V
90 ns at V_{DD} = 15.0 V

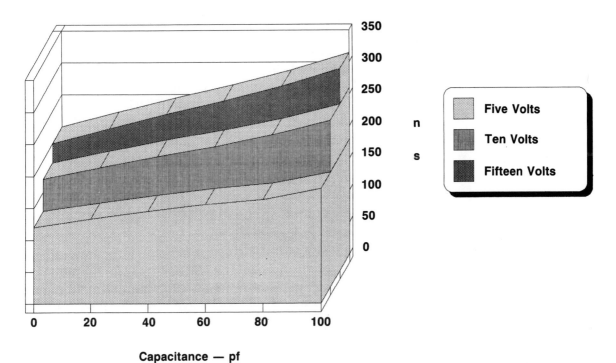

Capacitance — pf

Fig. 7-7. Propagation delay for CD4013.

CD4027 CMOS DUAL J-K
MASTER-SLAVE FLIP-FLOP

The CD4027 CMOS Dual J-K Master-Slave Flip-Flop consists of two J-K master-slave flip-flops. These flip-flops are completely static with their states remaining intact with a 1 or 0 logic clock. Each flip-flop possesses separate J, K, set, clock, and reset inputs, along with dual buffered outputs (one output is inverted). These inputs and outputs are compatible with the operation of the CD4013 Dual D-Type Flip-Flop. All inputs are high logic active with the set and reset inputs being independent of the clock. The internal logic of the CD4027 is determined through the logic of the inputs J and K. These input's logic are changed on the positive-transition of the clock pulse.

Package Configuration: 16-lead DIP

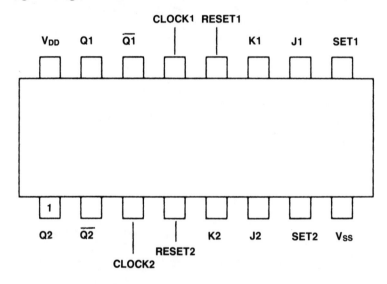

Fig. 7-8. Pin assignments for CD4027 CMOS Dual J-K Master-Slave Flip-Flop.

Maximum Input Capacitance: 7.5 pF

Clock Input Frequency: 3.5 - 24 MHz

Clock Pulse Width: 20 - 140 ns

Set Pulse Width: 25 - 180 ns

Reset Pulse Width: 25 - 180 ns

Data Setup Time: 25 - 200 ns

Maximum Propagation Delay Time; Clock-, Set-, or Reset-to-Outputs:
  380 ns at V_{DD} = 5.0 V
  130 ns at V_{DD} = 10.0 V
  90 ns at V_{DD} = 15.0 V

CD4042 CMOS QUAD CLOCKED 'D' LATCH

The CD4042 CMOS Quad Clocked 'D' Latch has an internal logic of four latches strobed by a common clock. Each latch possesses separate data inputs along with dual complementary buffered outputs. Input data are transferred to both outputs during a clock pulse transition. Whether this is a positive- or negative-transition pulse is determined by the logic on the POLARITY input. For example, a 1 logic on POLARITY input requires a negative-transition on the CLOCK input. The opposite is true of a 0 logic on the POLARITY input. These data are retained on the outputs until another transition occurs.

Package Configuration: 16-lead DIP

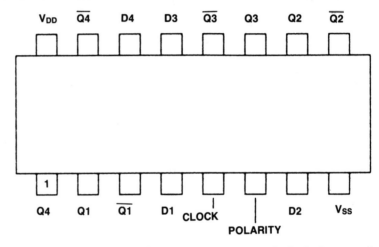

Fig. 7-9. Pin assignments for CD4042 CMOS Quad Clocked D Latch.

Maximum Input Capacitance: 7.5 pF

Clock Pulse Width: 30 - 200 ns

Hold Time: 25 - 120 ns

Data Setup Time: 0 - 50 ns

Maximum Propagation Delay Time; Data-to-Output:
220 ns at V_{DD} = 5.0 V
110 ns at V_{DD} = 10.0 V
80 ns at V_{DD} = 15.0 V

Maximum Propagation Delay Time; Clock-to-Output:
450 ns at V_{DD} = 5.0 V
200 ns at V_{DD} = 10.0 V
160 ns at V_{DD} = 15.0 V

CD4099 CMOS EIGHT-BIT ADDRESSABLE LATCH

The CD4099 CMOS Eight-Bit Addressable Latch is a programmable 8-bit serial input, parallel output decoder/latch device. The 3- bit decoder establishes the address for the input of the serial data into one of the eight latches. A WRITE DISABLE input governs the input of these data with a 0 logic enabling data input. These data are retained on the outputs until another data input occurs.

A RESET input provides a master reset capability that returns all output data bits to a 0 logic. This condition is met when a 0 logic on the RESET input is coupled with a 1 logic on the WRITE DISABLE input. Another operation mode is possible when the RESET input is at a 0 logic and the WRITE DISABLE input has a low logic. This condition makes the CD4099 act like a 1-of-8 demultiplexer. In this mode, the addressed output follows the data input while all other outputs are fixed at a 0 logic state.

Package Configuration: 16-lead DIP

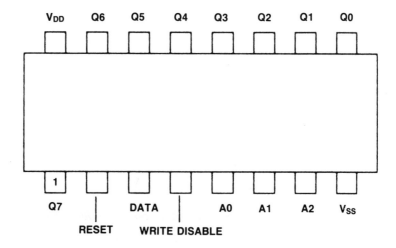

Fig. 7-10. Pin assignments for CD4099 CMOS Eight-Bit Addressable Latch.

Maximum Input Capacitance: 7.5 pF

Data Pulse Width: 40 - 200 ns

Address Pulse Width: 65 - 400 ns

Reset Pulse Width: 25 - 150 ns

Hold Time: 25 - 150 ns

Data Setup Time: 20 - 100 ns

Maximum Propagation Delay Time; Data-to-Output:
400 ns at V_{DD} = 5.0 V
150 ns at V_{DD} = 10.0 V
100 ns at V_{DD} = 15.0 V

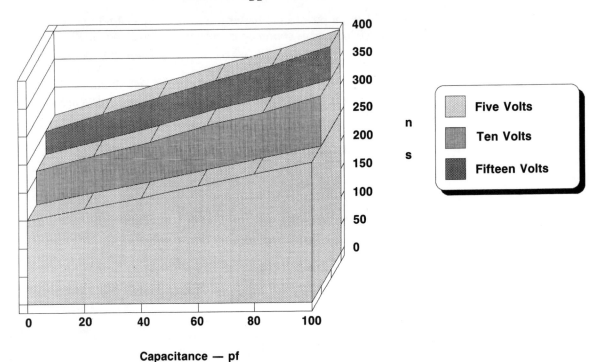

Capacitance — pf

Fig. 7-11. Propagation delay for CD4099.

Maximum Propagation Delay Time; Address-to-Output:
450 ns at V_{DD} = 5.0 V
200 ns at V_{DD} = 10.0 V
150 ns at V_{DD} = 15.0 V

CD4508 CMOS DUAL FOUR-BIT LATCH

The CD4508 CMOS Dual Four-Bit Latch consists of two 4-bit latches with individual inputs, controls, and outputs. The data on the four inputs are latched through three separate control inputs: OUTPUT DISABLE, STROBE, and RESET. Typically, the outputs will follow the logic of the data inputs, unless the OUTPUT DISABLE is at a high logic state. This condition forces the outputs into a high-impedance three-state condition. Otherwise, when a 0 logic is on the OUTPUT DISABLE input the STROBE input is used for latching the data on the corresponding output. A 0 logic on the STROBE input is used for latching these data on their outputs.

The RESET input provides a master reset capability that returns all output data bits to a 0 logic. This condition is met with a 0 logic on the RESET input regardless of the state of the STROBE input.

Package Configuration: 24-lead DIP

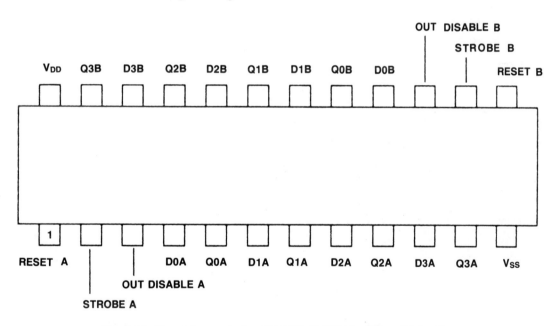

Fig. 7-12. Pin assignments for CD4508 CMOS Dual Four-Bit Latch.

Maximum Input Capacitance: 7.5 pF

Strobe Pulse Width: 35 - 140 ns

Reset Pulse Width: 50 - 200 ns

Data Setup Time: 10 - 50 ns

Maximum Propagation Delay Time; Data-to-Output:
210 ns at V_{DD} = 5.0 V
120 ns at V_{DD} = 10.0 V
90 ns at V_{DD} = 15.0 V

Maximum Propagation Delay Time; Strobe-to-Output:
260 ns at V_{DD} = 5.0 V
140 ns at V_{DD} = 10.0 V
100 ns at V_{DD} — 15.0 V

Maximum Propagation Delay Time; Reset-to-Output:
180 ns at V_{DD} = 5.0 V
100 ns at V_{DD} = 10.0 V
80 ns at V_{DD} = 15.0 V

Maximum Propagation Delay Time; Three-State:
180 ns at V_{DD} = 5.0 V
100 ns at V_{DD} = 10.0 V
70 ns at V_{DD} = 15.0 V

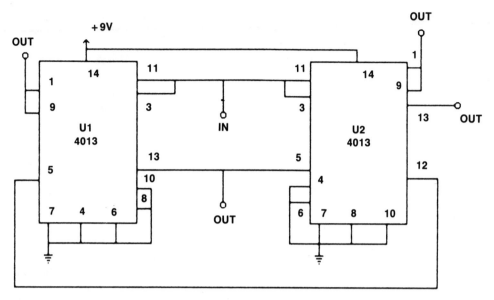

Fig. 7-13. Schematic diagram for Binary Counter.

BINARY COUNTER

Several simple counting circuits can be built around the versatile CD4013 Dual D -Type Flip-Flop (see Figs. 7-13, 7-14, and 7-15). Each of these projects uses one (or two) CD4013s for changing their internal logic states in response to a clock pulse. Depending on the configuration of the output, a binary count,

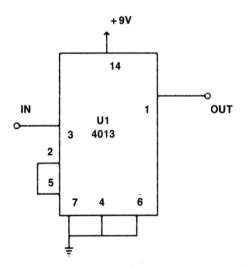

Fig. 7-14. Schematic diagram for Pulse-Divided Counter.

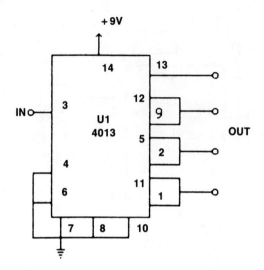

Fig. 7-15. Schematic diagram for Sequential Counter.

pulse-divided count, or a sequential count can be executed. An ideal clock circuit for driving these counters could be designed from the CD4011 Quad NAND Gate.

In the case of the Binary Counter, two CD4013's are linked together through the second-stage output of one flip-flop, and the first-stage data input of the second flip-flop. This arrangement can be further expanded for providing large 8-, 12-, and 16-bit counters.

8

Decoders

CD4028 CMOS BCD-TO-DECIMAL DECODER

The CD4028 CMOS BCD-to-Decimal Decoder contains four buffered inputs driving 10 buffered outputs. A battery of internal decoding logic gates can be configured for either BCD-to-decimal output or binary-to-octal output. The configuration of this output is determined through the nature of the inputs. If a BCD code input is applied to all four inputs (A, B, C, and D), then the selected output will have a decimal-coded high logic. Only the decimal digits 0 through 9 can be output from this mode. On the other hand, if a three-bit code is applied to three inputs (A, B, and C with input D having a 0 logic), then the resultant output will have an octal-coded high logic.

Package Configuration: 16-lead DIP

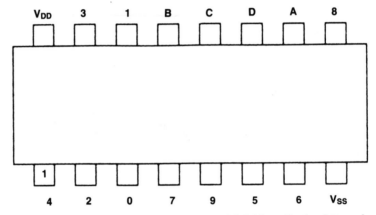

Fig. 8-1. Pin assignments for CD4028 CMOS BCD-to-Decimal Decoder.

Maximum Input Voltage; Low Level: 1.5 V at V_{DD} = 5.0 V
3.0 V at V_{DD} = 10.0 V
4.0 V at V_{DD} = 15.0 V

Maximum Output Voltage; Low Level: 0.05 V

Maximum Propagation Delay Time: 350 ns at V_{DD} = 5.0 V
160 ns at V_{DD} = 10.0 V
120 ns at V_{DD} = 15.0 V

CD4514 CMOS FOUR-BIT
LATCH/4-TO-16 LINE DECODER

The CD4514 CMOS Four-Bit Latch/4-to-16 Line Decoder with outputs high on select contains a four-bit strobed latch connected to a 4-to-16 decoder. Data entered on the inputs is latched on the negative-transition of the STROBE input. Once latched, these 4-bit data are passed into a decoder for selectively enabling a corresponding output. The decimal digits 0 through 15 can be output in this manner. An INHIBIT control input is able to reset all outputs to a 0 logic, regardless of the latched data inputs.

Package Configuration: 24-lead DIP

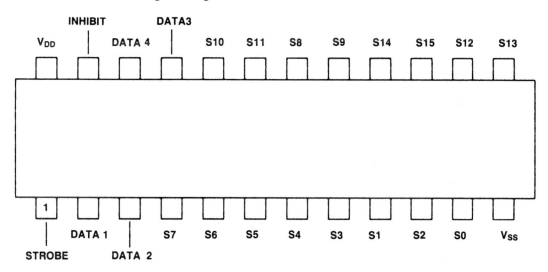

Fig. 8-2. Pin assignments for CD4514 CMOS Four-Bit Latch/4-to-16 Line Decoder with outputs high on select.

Data Setup Time: 20 - 150 ns

Strobe Pulse Width: 40 - 250 ns

Maximum Input Voltage; Low Level: 1.5 V at V_{DD} = 5.0 V
3.0 V at V_{DD} = 10.0 V
4.0 V at V_{DD} = 15.0 V

Maximum Output Voltage; Low Level: 0.05 V

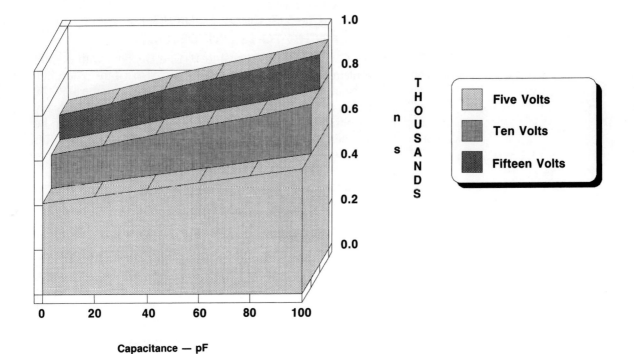

Fig. 8-3. Propagation delay for CD4514.

Maximum Propagation Delay Time; Data and Strobe:

970 ns at V_{DD} = 5.0 V
370 ns at V_{DD} = 10.0 V
270 ns at V_{DD} = 15.0 V

Maximum Propagation Delay Time; Inhibit:

500 ns at V_{DD} = 5.0 V
220 ns at V_{DD} = 10.0 V
170 ns at V_{DD} = 15.0 V

CD4515 CMOS FOUR-BIT
LATCH/4-TO-16 LINE DECODER

The CD4515 CMOS Four-Bit Latch/4-to-16 Line Decoder with outputs low on select contains a four-bit strobed latch connected to a 4-to-16 decoder. Data entered on the inputs are latched on the negative-transition of the STROBE input. Once latched, these 4-bit data are passed into a decoder for selectively enabling a corresponding output. The decimal digits 0 through 15 can be output in this manner. An INHIBIT control input is able to reset all outputs to a 1 logic, regardless of the latched data inputs.

Package Configuration: 24-lead DIP

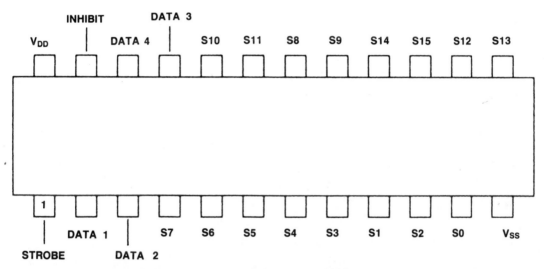

Fig. 8-4. Pin assignments for CD4515 CMOS Four-Bit Latch/4-to-16 Line Decoder with Outputs Low on Select.

Data Setup Time: 20 - 150 ns

Strobe Pulse Width: 40 - 250 ns

Maximum Input Voltage; Low Level: 1.5 V at V_{DD} = 5.0 V
3.0 V at V_{DD} = 10.0 V
4.0 V at V_{DD} = 15.0 V

Maximum Output Voltage; Low Level: 0.05 V

Maximum Propagation Delay Time; Data and Strobe:
970 ns at $V_{DD} = 5.0$ V
370 ns at $V_{DD} = 10.0$ V
270 ns at $V_{DD} = 15.0$ V

Maximum Propagation Delay Time; Inhibit:
500 ns at $V_{DD} = 5.0$ V
220 ns at $V_{DD} = 10.0$ V
170 ns at $V_{DD} = 15.0$ V

CD4555 CMOS DUAL BINARY-TO-1-OF-4 DECODER/DEMULTIPLEXER

The CD4555 CMOS Dual Binary-to-1-of-4 Decoder/Demultiplexer with outputs high on select has an internal logic of two 1-of-4 decoder/demultiplexers each with separate inputs and outputs. There are two select inputs (A and B) along with one active low logic enable input (E) coupled to four outputs for each decoder circuit. Data entered on the select inputs are present on one select output at a time. The enable input is used for resetting all of the outputs to a 0 logic, regardless of each select input's state.

Package Configuration: 16-lead DIP

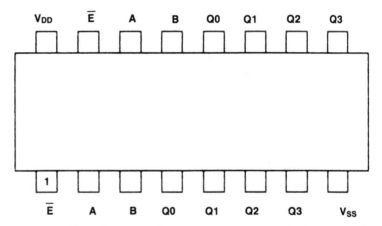

Fig. 8-5. Pin assignments for CD4555 CMOS Dual Binary-to-1-of-4 Decoder/Demultiplexer with Outputs High on Select.

Maximum Input Voltage; Low Level:
1.5 V at V_{DD} = 5.0 V
3.0 V at V_{DD} = 10.0 V
4.0 V at V_{DD} = 15.0 V

Maximum Output Voltage; Low Level: 0.05 V

Maximum Propagation Delay Time; Input-to-Output:
440 ns at V_{DD} = 5.0 V
190 ns at V_{DD} = 10.0 V
140 ns at V_{DD} = 15.0 V

Maximum Propagation Delay Time; Enable-to-Output:
400 ns at V_{DD} = 5.0 V
170 ns at V_{DD} = 10.0 V
130 ns at V_{DD} = 15.0 V

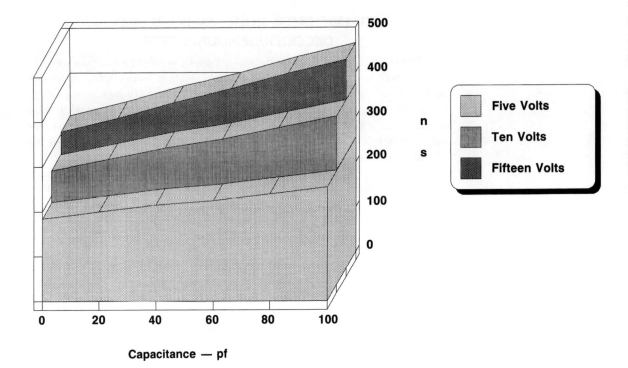

Fig. 8-6. Propagation delay for CD4555.

CD4556 CMOS DUAL BINARY-TO-1-OF-4 DECODER/DEMULTIPLEXER

The CD4556 CMOS Dual Binary-to-1-of-4 Decoder/Demultiplexer with outputs low on select has an internal logic of two 1-of-4 decoder/demultiplexers each with separate inputs and outputs. There are two select inputs (A and B) along with one active low logic enable input (E) coupled to four inverted outputs for each decoder circuit. Data entered on the select inputs are present on one select output at a time. The enable input is used for resetting all of the outputs to a 1 logic, regardless of each select input's state.

Package Configuration: 16-lead DIP

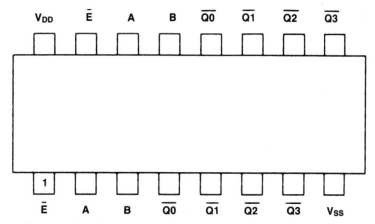

Fig. 8-7. Pin assignments for CD4556 CMOS Dual Binary-to-1-of-4 Decoder/Demultiplexer with Outputs Low on Select.

Maximum Input Voltage; Low Level: 1.5 V at V_{DD} = 5.0 V
3.0 V at V_{DD} = 10.0 V
4.0 V at V_{DD} = 15.0 V

Maximum Output Voltage; Low Level: 0.05 V

Maximum Propagation Delay Time; Input-to-Output:
440 ns at V_{DD} = 5.0 V
190 ns at V_{DD} = 10.0 V
140 ns at V_{DD} = 15.0 V

Maximum Propagation Delay Time; Enable-to-Output:
400 ns at V_{DD} = 5.0 V
170 ns at V_{DD} = 10.0 V
130 ns at V_{DD} = 15.0 V

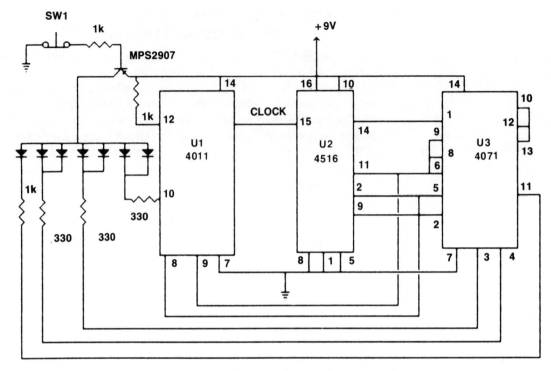

Fig. 8-8. Schematic diagram for CMOS Die.

CMOS DIE

An interesting challenge to the CMOS designer is the construction of a simple circuit that duplicates the numeric results of a die. One possible solution to this dilemma is illustrated in Fig. 8-8. In this example, the die pips are simulated with seven red LEDs.

Basically, the CMOS die performs its 1 through 6 number selection in three different stages of operation. First, the CD4011 is used as a clock circuit for driving the counting pulses of the CD4516. In turn, the output from the CD4516 is directed through various gate combinations in the CD4071, along with a final NAND gate from the CD4011. The result is the pseudo-random lighting of the die's pips.

In theory, the CMOS die begins its display of a number with the activation of switch SW2. This pushbutton not only initiates the clock pulse, it also disables the LEDs. Therefore, while the CMOS die is busy generating a number output, the die "face" is blank. As the clock pulses are sent to the CD4516, the counter's outputs cycle between 0 and 7 until the clock signal is halted with the deactivation of SW2. At this time, the selected LEDs are lighted in a die representation. The proper lighting sequence for the LEDs is determined by the logic on the OR and NAND gates connected to the CD4516's outputs.

9

High-Speed CMOS Technology

74HC00 QUAD TWO-INPUT NAND

The 74HC00 Quad Two-Input NAND gate has an internal logic of four two-input NAND gates. All inputs are buffered with the outputs able to drive ten LSTTL loads at near LSTTL processing speeds.

Package Configuration: 14-lead DIP

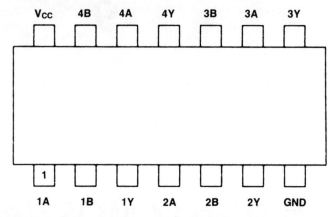

Fig. 9-1. Pin assignments for 74HC00 Quad Two-Input NAND Gate.

Power Dissipation Capacitance: 25 pF

Maximum Input Voltage; Low-Level: 0.5 V at V_{CC} = 2.0 V
$\qquad\qquad\qquad\qquad\qquad\qquad\qquad$ 1.35 V at V_{CC} = 4.5 V
$\qquad\qquad\qquad\qquad\qquad\qquad\qquad$ 1.8 V at V_{CC} = 6.0 V

Maximum Propagation Delay Time; Input-to-Output:
$\quad$ 115 ns at V_{CC} = 2.0 V
$\quad$ 23 ns at V_{CC} = 4.5 V
$\quad$ 20 ns at V_{CC} = 6.0 V

Maximum Transition Time: 95 ns at V_{CC} = 2.0 V
$\qquad\qquad\qquad\qquad\qquad\quad$ 19 ns at V_{CC} = 4.5 V
$\qquad\qquad\qquad\qquad\qquad\quad$ 16 ns at V_{CC} = 6.0 V

74HC08 QUAD TWO-INPUT AND

The 74HC08 Quad Two-Input AND gate has an internal logic of four two-input AND gates. All inputs are buffered with the outputs able to drive ten LSTTL loads at near LSTTL processing speeds.

Package Configuration: 14-lead DIP

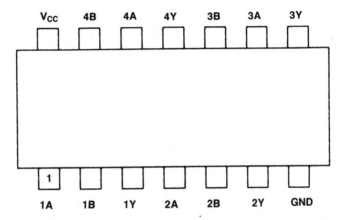

Fig. 9-2. Pin assignments for 74HC08 Quad Two-Input AND Gate.

Power Dissipation Capacitance: 37 pF

Maximum Input Voltage; Low-Level:
    0.5 V at V_{CC} = 2.0 V
    1.35 V at V_{CC} = 4.5 V
    1.8 V at V_{CC} = 6.0 V

Maximum Propagation Delay Time; Input-to-Output:
    115 ns at V_{CC} = 2.0 V
    23 ns at V_{CC} = 4.5 V
    20 ns at V_{CC} = 6.0 V

Maximum Transition Time:
    95 ns at V_{CC} = 2.0 V
    19 ns at V_{CC} = 4.5 V
    16 ns at V_{CC} = 6.0 V

74HC27 TRIPLE THREE-INPUT NOR

The 74HC27 Triple Three-Input NOR gate has an internal logic of three three-input NOR gates. All inputs are buffered with the outputs able to drive ten LSTTL loads at near LSTTL processing speeds.

Package Configuration: 14-lead DIP

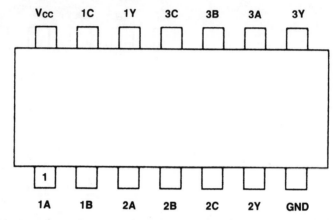

Fig. 9-3. Pin assignments for 74HC27 Triple Three-Input NOR Gate.

Power Dissipation Capacitance: 26 pF

Maximum Input Voltage; Low-Level:

   0.5 V at V_{CC} = 2.0 V
   1.35 V at V_{CC} = 4.5 V
   1.8 V at V_{CC} = 6.0 V

Maximum Propagation Delay Time; Input-to-Output:

   125 ns at V_{CC} = 2.0 V
   25 ns at V_{CC} = 4.5 V
   21 ns at V_{CC} = 6.0 V

Maximum Transition Time:

   95 ns at V_{CC} = 2.0 V
   19 ns at V_{CC} = 4.5 V
   16 ns at V_{CC} = 6.0 V

74HC32 QUAD TWO-INPUT OR

The 74HC32 Quad Two-Input OR gate is arranged with an internal logic of four two-input OR gates. The outputs are able to drive ten LSTTL loads at near LSTTL processing speeds.

Package Configuration: 14-lead DIP

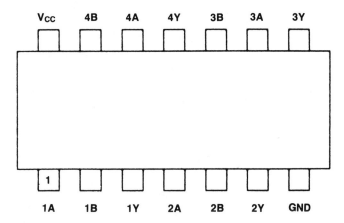

Fig. 9-4. Pin assignments for 74HC32 Quad Two-Input OR Gate.

Power Dissipation Capacitance: 22 pF

Maximum Input Voltage; Low-Level:

0.5 V at V_{CC} = 2.0 V
1.35 V at V_{CC} = 4.5 V
1.8 V at V_{CC} = 6.0 V

Maximum Propagation Delay Time; Input-to-Output:

115 ns at V_{CC} = 2.0 V
23 ns at V_{CC} = 4.5 V
20 ns at V_{CC} = 6.0 V

Maximum Transition Time:

95 ns at V_{CC} = 2.0 V
19 ns at V_{CC} = 4.5 V
16 ns at V_{CC} = 6.0 V

74HC42 BCD-TO-DECIMAL DECODER

The 74HC42 BCD-to-Decimal Decoder (1-of-10) has one of its ten outputs driven high by the BCD data on the inputs. If non-BCD data is present on the inputs, then all outputs are placed in a high logic state. The outputs are able to drive ten LSTTL loads at near LSTTL processing speeds.

Package Configuration: 16-lead DIP

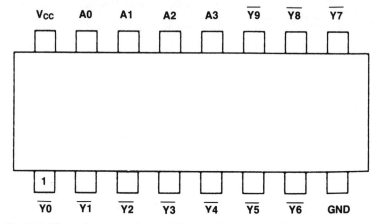

Fig. 9-5. Pin assignments for 74HC42 BCD-to-Decimal Decoder (1-of-10).

Power Dissipation Capacitance: 65 pF

Maximum Input Voltage; Low-Level:
    0.5 V at V_{CC} = 2.0 V
    1.35 V at V_{CC} = 4.5 V
    1.8 V at V_{CC} = 6.0 V

Maximum Propagation Delay Time; Input-to-Output:
    190 ns at V_{CC} = 2.0 V
    38 ns at V_{CC} = 4.5 V
    33 ns at V_{CC} = 6.0 V

Maximum Transition Time:
    95 ns at V_{CC} = 2.0 V
    19 ns at V_{CC} = 4.5 V
    16 ns at V_{CC} = 6.0 V

74HC74 DUAL "D"-TYPE FLIP-FLOP

The 74HC74 Dual 'D'-Type Flip-Flop with Set and Reset has two independent flip-flops with separate buffered data and control inputs, and complementary outputs. A data input is placed on the input on the positive-transition of the clock signal.

Package Configuration: 14-lead DIP

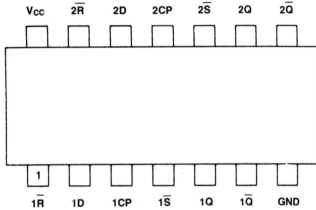

Fig. 9-6. Pin assignments for 74HC74 Dual 'D'-Type Flip-Flop with Set and Reset.

Power Dissipation Capacitance: 25 pF

Clock Frequency: 50 MHz

Maximum Input Voltage; Low-Level:
    0.5 V at V_{CC} = 2.0 V
    1.35 V at V_{CC} = 4.5 V
    1.8 V at V_{CC} = 6.0 V

Maximum Propagation Delay Time; Clock-to-Output:
    220 ns at V_{CC} = 2.0 V
    44 ns at V_{CC} = 4.5 V
    37 ns at V_{CC} = 6.0 V

Maximum Propagation Delay Time; Reset- and Set-to-Output:
    250 ns at V_{CC} = 2.0 V
    50 ns at V_{CC} = 4.5 V
    43 ns at V_{CC} = 6.0 V

Maximum Transition Time:
    95 ns at VCC = 2.0 V
    19 ns at VCC = 4.5 V
    16 ns at VCC = 6.0 V

74HC151 EIGHT-INPUT MULTIPLEXER

The 74HC151 Eight-Input Multiplexer has one 8-channel multiplexer with eight buffered data inputs, three buffered control inputs, and two complementary outputs. The outputs are able to drive ten LSTTL loads at near LSTTL processing speeds.

Package Configuration: 16-lead DIP

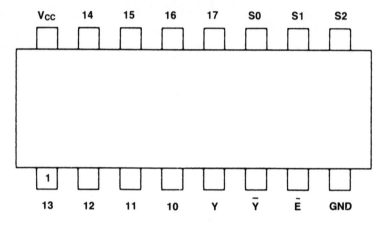

Fig. 9-7. Pin assignments for 74HC151 Eight-Input Multiplexer.

Power Dissipation Capacitance: 59 pF

Maximum Input Voltage; Low-Level:
0.5 V at V_{CC} = 2.0 V
1.35 V at V_{CC} = 4.5 V
1.8 V at V_{CC} = 6.0 V

Maximum Propagation Delay Time; Data-to-Output:
215 ns at V_{CC} = 2.0 V
43 ns at V_{CC} = 4.5 V
37 ns at V_{CC} = 6.0 V

Maximum Propagation Delay Time; Select-to-Output:
230 ns at V_{CC} = 2.0 V
46 ns at V_{CC} = 4.5 V
39 ns at V_{CC} = 6.0 V

Maximum Propagation Delay Time; Enable-to-Output:

175 ns at V_{CC} = 2.0 V

35 ns at V_{CC} = 4.5 V

30 ns at V_{CC} = 6.0 V

Maximum Transition Time:

95 ns at V_{CC} = 2.0 V

19 ns at V_{CC} = 4.5 V

16 ns at V_{CC} = 6.0 V

74HC164 EIGHT-BIT
SERIAL-IN/PARALLEL-OUT SHIFT REGISTER

The 74HC164 Eight-Bit Serial-In/Parallel-Out Shift Register has an SI/PO with an asynchronous master reset. There are two buffered serial data inputs, and two buffered control inputs. Data are shifted during a positive-transition clock pulse.

Package Configuration: 14-lead DIP

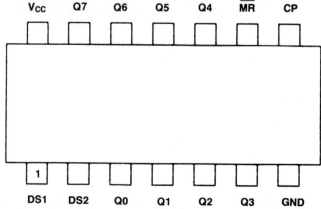

Fig. 9-8. Pin assignments for 74HC164 Eight-Bit Serial-In/Parallel-Out Shift Register.

Power Dissipation Capacitance: 47 pF

Clock Frequency: 5 - 28 MHz

Maximum Input Voltage; Low-Level:

0.5 V at V_{CC} = 2.0 V
1.35 V at V_{CC} = 4.5 V
1.8 V at V_{CC} = 6.0 V

Maximum Propagation Delay Time; Clock-to-Output:

212 ns at V_{CC} = 2.0 V
43 ns at V_{CC} = 4.5 V
36 ns at V_{CC} = 6.0 V

Maximum Propagation Delay Time; Master Reset-to-Output:

200 ns at V_{CC} = 2.0 V
40 ns at V_{CC} = 4.5 V
34 ns at V_{CC} = 6.0 V

Maximum Transition Time:

95 ns at V_{CC} = 2.0 V
19 ns at V_{CC} = 4.5 V
16 ns at V_{CC} = 6.0 V

74HC243 QUAD-BUS TRANSCEIVER

The 74HC243 Quad-Bus Transceiver with Three-State Outputs provides two-way asynchronous data communication. There are four bi-directional buffered non-inverting inputs, and four 3-state outputs. Both the data direction and the output mode are determined by the logic of the output enable inputs. The outputs are able to drive ten LSTTL loads at near LSTTL processing speeds.

Package Configuration: 14-lead DIP

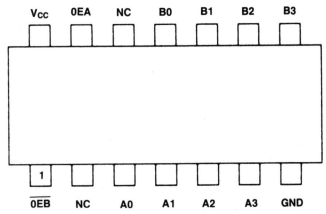

Fig. 9-9. Pin assignments for 74HC243 Quad-Bus Transceiver with Three-State Outputs.

Power Dissipation Capacitance: 66 pF

Maximum Input Voltage; Low-Level:
0.5 V at V_{CC} = 2.0 V
1.35 V at V_{CC} = 4.5 V
1.8 V at V_{CC} = 6.0 V

Maximum Propagation Delay Time; Data-to-Output:
140 ns at V_{CC} = 2.0 V
28 ns at V_{CC} = 4.5 V
24 ns at V_{CC} = 6.0 V

Maximum Propagation Delay Time; Output Z-to-Output Low:
250 ns at V_{CC} = 2.0 V
50 ns at V_{CC} = 4.5 V
43 ns at V_{CC} = 6.0 V

Maximum Transition Time:
75 ns at V_{CC} = 2.0 V
15 ns at V_{CC} = 4.5 V
13 ns at V_{CC} = 6.0 V

74HC688 EIGHT-BIT MAGNITUDE COMPARATOR

The 74HC688 Eight-Bit Magnitude Comparator is able to compare the logic of two separate 8-bit words. The single output is a low logic only when the two byte-sized words being compared are equal. The outputs are able to drive ten LSTTL loads at near LSTTL processing speeds.

Package Configuration: 20-lead DIP

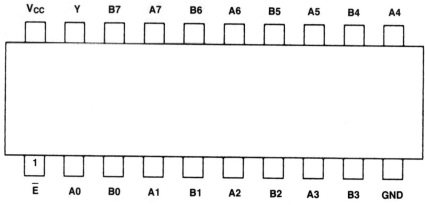

Fig. 9-10. Pin assignments for 74HC688 Eight-Bit Magnitude Comparator.

Power Dissipation Capacitance: 22 pF

Maximum Input Voltage; Low-Level:
   0.5 V at V_{CC} = 2.0 V
   1.35 V at V_{CC} = 4.5 V
   1.8 V at V_{CC} = 6.0 V

Maximum Propagation Delay Time; Data-to-Output:
   212 ns at V_{CC} = 2.0 V
   42 ns at V_{CC} = 4.5 V
   36 ns at V_{CC} = 6.0 V

Maximum Propagation Delay Time; Enable-to-Output:
   150 ns at V_{CC} = 2.0 V
   30 ns at V_{CC} = 4.5 V
   26 ns at V_{CC} = 6.0 V

Maximum Transition Time:
   95 ns at V_{CC} = 2.0 V
   19 ns at V_{CC} = 4.5 V
   16 ns at V_{CC} = 6.0 V

8K RAM CARD

Many of the popular notebook-sized microcomputers that are available on today's marketplace use CMOS circuitry throughout their design. This construction technique enables the computer to be operated with a minimal battery power supply for extended periods of time. Two popular examples of this genre of computer are the Tandy Model 100 and the NEC PC-8201A (Note: both computers were actually designed by Kyocera of Japan). Both of these computers not only share a similar external appearance, but they also have the same basic internal specifications.

A powerful CMOS version of the 8085 MPU, the 80C85, supplies the main control circuitry for both of these computers. Several other CMOS LSI (Large-Scale Integration) devices, including an 81C55RS PIO, a 1990AC Timer, 5518 RAM, and 535618 ROM, support the 80C85. While access to these

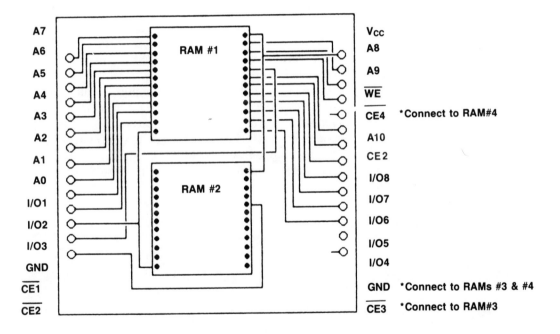

NOTE: RAM #3 and RAM #4 are located on the
bottom of the board carrier. Use CE3 and CE4
for connecting CE1 of these two RAMs, respectively.

USE TOSHIBA TC5518BBF-25 OR 6118LP-4 (200 NS) RAM
CHIPS.

Fig. 9-11. Schematic diagram for 8K RAM Card.

components is limited, there is a RAM expansion area in both computers that is able to support the addition of supplemental RAM chips.

RAM expansion in either the Tandy Model 100 or the NEC PC-8201A is accomplished with 8K × 8 RAM cards. Each computer is able to hold the installation of three 8K RAM cards (six 8K RAM cards can be inserted into the NEC PC-8201A). Figure 9-11 is a schematic diagram for wiring the 8K RAM Card. This memory expansion card must be constructed from four 16K-bit or 2048 × 8 TC5518BF-25 SMT (Surface Mount Technology) Static RAM chips. Based on the wiring in Fig. 9-11 the 24-pin 5518s can be easily added to the memory map of the notebook computer.

In this configuration the individual 5518s are selected through the active low logic, twin chip enable inputs. Based on the design of the 8K RAM Card, a fully configured Tandy Model 100 computer will have a total of 16 5518 static RAM chips. Therefore, address decoding and bank selection will require 16 chip-select signals.

The RAM memory map for the Model 100 typically runs from 8000 to FFFF (hex). Addressing this area is provided by three control signal inputs, and three chip-select inputs. Two digital CMOS ICs provide the decoding logic for generating these inputs. A Dual 2-to-4 Line Decoder/Demultiplexer (TC40H139) produces the three control inputs, while two 3-to-8 Line Decoder/Demultiplexers (TC40H138) create the 16 chip-select signals from the three chip-select inputs. Interestingly enough, the same IC that generates the three control inputs, (TC40H139) also controls the chip-select signals used by the notebook computer's CMOS ROM. This ROM portion is located from 0000 to 7FFF (hex) in the memory scheme of these Kyocera twins.

10

Memory and LSI Circuitry

5101 CMOS
256-WORD-BY-FOUR-BIT LSI STATIC RAM

The 5101 CMOS 256-Word-by-Four-Bit LSI Static RAM is a static memory device with an internal organization of 256 × 4 bits. Eight address lines are controlled with four inputs for accessing four data input and four data output lines. These four control inputs consist of two complementary chip select inputs, a read/write mode select input, and an output disable input. The use of the output disable forces all outputs into a high-impedance 3-state output.

Package Configuration: 22-lead DIP

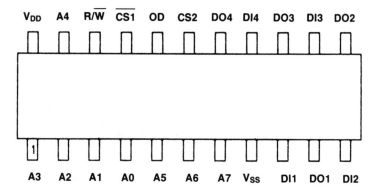

Fig. 10-1. Pin assignments for 5101 CMOS 256-Word-by-Four-Bit LSI Static RAM.

Output Capacitance: 10 pF

Data Retention Voltage: 1.5 - 2.0 V

Rise and Fall Time: 1 ms

Read Times: Read Cycle: 250 - 350 ns
Address Access: 150 - 350 ns

Write Times: Write Cycle: 300 - 400 ns
Address Setup: 110 - 150 ns

6116 CMOS
2048-WORD-BY-EIGHT-BIT STATIC RAM

The 6116 CMOS 2048-Word-by-Eight-Bit Static RAM is a static memory device with an internal organization of 2048 × 8 bits. Eight address lines are controlled with three inputs for accessing eight dual data input and output ports. These three active low logic control inputs consist of a chip enable input, a write enable input, and an output enable input. The presence of the chip enable inputs, and the output enable inputs allows for the employment of this device in memory expansion.

Package Configuration: 24-lead DIP

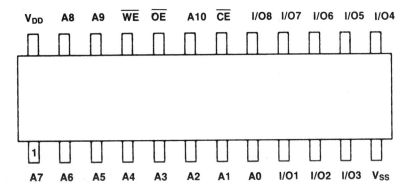

Fig. 10-2. Pin assignments for 6116 CMOS 2048-Word-by-Eight-Bit Static RAM.

Output Capacitance: 6 pF

Data Retention Voltage: 2.0 - 4.5 V

Read Times: Read Cycle: 150 - 250 ns
Address Access: 150 - 250 ns

Write Times: Write Cycle: 150 - 250 ns
Address Setup: 0 ns
Write Pulse Width: 90 - 200 ns

6264 CMOS
8192-WORD-BY-EIGHT-BIT LSI STATIC RAM

The 6264 CMOS 8192-Word-by-Eight-Bit LSI Static RAM is a static memory device with an internal organization of 8192 × 8 bits. Eleven address lines are controlled with four inputs for accessing eight dual data input and output ports. These four control inputs include a pair of complementary chip enable inputs, an active low logic write enable input, and an active low logic output enable input.

Package Configuration: 28-lead DIP

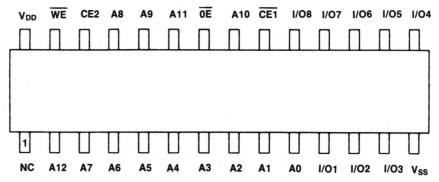

Fig. 10-3. Pin assignments for 6264 CMOS 8192-Word-by-Eight-Bit Static RAM.

Output Capacitance: 6 pF

Data Retention Voltage: 2.0 - 5.5 V

Read Times: Read Cycle: 120 - 150 ns
Address Access: 120 - 150 ns

Write Times: Write Cycle: 120 - 150 ns
Address Setup: 0 ns
Write Enable Width: 80 - 100 ns

27C64 CMOS 65,536-BIT UV ERASABLE AND ELECTRICALLY PROGRAMMABLE READ ONLY MEMORY

The 27C64 CMOS 65,536-Bit UV Erasable and Electrically Programmable Read Only Memory is a static memory device with an internal organization of 8192 × 8 bits. Thirteen address lines are controlled with two inputs for accessing eight dual data input and output ports. These two active low logic control inputs include a chip enable input, and an output enable input. Two other control inputs are used only during device data programming. A transparent opening in the dorsal surface of the device permits the erasure of all data through the application of UV light.

Package Configuration: 28-lead DIP

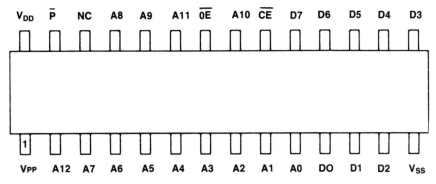

Fig. 10-4. Pin assignments for 27C64 CMOS 65,536-Bit EPROM.

Output Capacitance: 8 pF

Read Times: Read Cycle: 200 - 300 ns
Data Setup: 2 ms

Write Times: Write Cycle: 200 - 300 ns
Address Setup: 2 ms
Output Enable Setup: 2 ms

27C128 CMOS 131,072-BIT UV EEPROM

The 27C128 CMOS 131,072-Bit UV EEPROM is a static memory device with an internal organization of 16384 × 8 bits. Fourteen address lines are controlled with two inputs for accessing eight dual data input and output ports. These two active low logic control inputs include a chip enable input, and an output enable input. Two other control inputs are used only during device data programming. A transparent opening in the dorsal surface of the device permits the erasure of all data through the application of UV light.

Package Configuration: 28-lead DIP

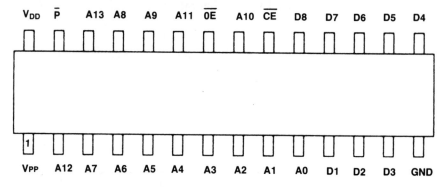

Fig. 10-5. Pin assignments for 27C128 CMOS 131,072-Bit EPROM.

Output Capacitance: 8 pF

Read Times: Read Cycle: 200 - 250 ns
Data Setup: 2 ms

Write Times: Write Cycle: 200 - 250 ns
Address Setup: 2 ms
Output Enable Setup: 2 ms

27C256 CMOS 262,144-BIT UV EEPROM

The 27C256 CMOS 262,144-Bit UV EEPROM is a static memory device with an internal organization of 32768 × 8 bits. Fifteen address lines are controlled with two inputs for accessing eight dual data input and output ports. These two active low logic control inputs include a chip enable input, and an output enable input. Two other control inputs are used only during device data programming. A transparent opening in the dorsal surface of the device permits the erasure of all data through the application of UV light.

Package Configuration: 28-lead DIP

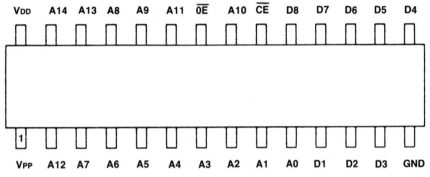

Fig. 10-6. Pin assignments for 27C256 CMOS 262,144-Bit EPROM.

Output Capacitance: 8 pF

Read Times: Read Cycle: 250 - 300 ns
Data Setup: 2 ms

Write Times: Write Cycle: 250 - 300 ns
Address Setup: 2 ms
Output Enable Setup: 2 ms

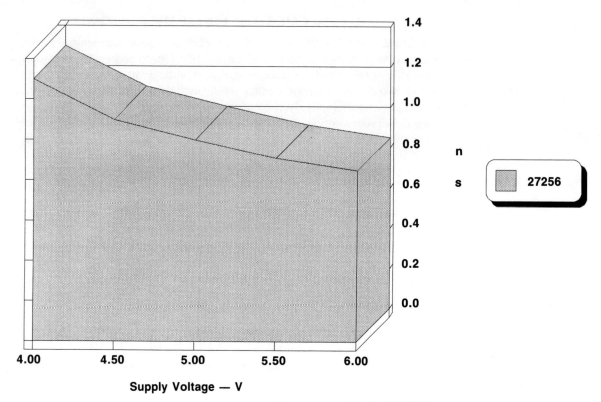

Fig. 10-7. Address access time for 27C256.

27C512 CMOS 524,288-BIT UV EEPROM

The 27C512 CMOS 524,288-Bit UV EEPROM is a static memory device with an internal organization of 65536 × 8 bits. Sixteen address lines are controlled with two inputs for accessing eight dual data input and output ports. These two active low logic control inputs include a chip enable input and an output enable input (this input also doubles at the programming voltage pin). Two other control inputs are used only during device data programming. A transparent opening in the dorsal surface of the device permits the erasure of all data through the application of UV light.

Package Configuration: 28-lead DIP

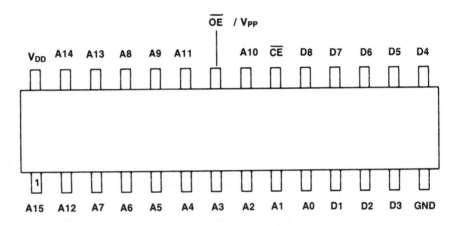

Fig. 10-8. Pin assignments for 27C512 CMOS 524,288-Bit EPROM.

Output Capacitance: 8 pF

Read Times: Read Cycle: 250 - 300 ns
Data Setup: 2 ms

Write Times: Write Cycle: 250 - 300 ns
Address Setup: 2 ms
Output Enable Setup: 2 ms

27C1028 CMOS 1,048,576-BIT UV EEPROM

The 27C1028 CMOS 1,048,576-Bit UV EEPROM is a static memory device with an internal organization of 65536 × 16 bits. Seventeen address lines are controlled with five inputs for accessing sixteen, dual data/address input and output ports. These five active low logic control inputs include a chip enable input, an output enable input, address logic enable, program logic, and electrical enable control signal input. The program logic input, and another control input are used only during device data programming. A transparent opening in the dorsal surface of the device permits the erasure of all data through the application of UV light.

Package Configuration: 28-lead DIP

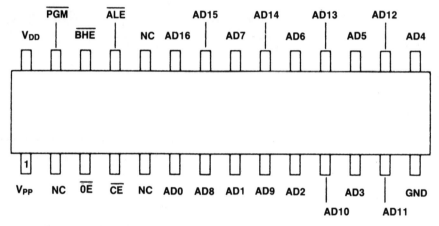

Fig. 10-9. Pin assignments for 27C1028 CMOS 1,048,576-Bit EPROM.

Output Capacitance: 8 pF

Read Times: Read Cycle: 200 - 250 ns
Data Setup: 2 ms

Write Times: Write Cycle: 200 - 250 ns
Address Setup: 1 ms
Output Enable Setup: 2 ms

TEXT-TO-SPEECH SYNTHESIZER

In 1976, the Naval Research Laboratories developed a small and compact algorithm that could handle all of the analyzing and translation duties that are normally associated with converting a written word into a speech synthesized word. The elegant nature of this text-to-speech algorithm has made it the most popular method for converting ASCII text into synthesized speech.

The General Instrument CTS256A-AL2 is a NMOS IC that contains an on-board text-to-speech algorithm that translates ASCII text into allophonic code strings. These allophonic code strings are specifically matched to the data input signals that are used by the General Instrument SPO256-AL2 speech synthesizer IC. Therefore, tying the CTS256A-AL2 together with the SPO256-AL2 creates an inexpensive text-to-speech processing system.

Figure 10-10 is a serial interface text-to-speech ASCII translation speech synthesizer circuit. One problem that is associated with this type of speech synthesizer is that it is incapable of correctly pronouncing certain words (e.g., joking). In an effort to hurdle this pronunciation difficulty, special ROM programming can be used as a "look-up" data table for holding the correct pronunciation for a given ASCII text spelling. In other words, this look-up table would contain the required phonemes for correctly uttering the above example word, "joking." Later when the ASCII text for "joking" is encountered by the CTS256A-AL2, the data table will provide the correct pronunciation.

In this circuit, an EPROM is used for holding the look-up data table pronunciations. A programmed 27C64 EPROM, IC8 in this diagram, serves as the site for this look-up phoneme data table. Inside this EPROM, the data starts at an address of 3000h. As an example of programming this EPROM, here is the procedure that would be used for coding the pronunciation of "joking:"

<[JOKING]< = [JH OW KK1 IH NG]
13 6F 2B 29 2E A7 4A 35 2A 0C AC

Programming the EPROM with "joking:"

```
3000 80 48 28 58 85 E0 35 E0 31 FF FF FF FF FF FF FF
3010 FF FF FF FF FF FF FF FF FF FF FF FF FF FF FF FF
3020 FF FF FF 1E 1F 20 21 28 29 24 25 22 23 2A 2B 26
     ; INSERT MAIN CONTROL PROGRAM
30A0 00 E0 36 31 93 31 AB 31 A9 31 B1 31 B2 31 B3 31
30B0 B4 31 E1 31 E2 32 0D 32 0E 32 0F 32 1B 32 1C 32
30C0 1D 32 1E 32 2D 32 2E 32 2F 32 30 32 3D 32 5A 32
30D0 5B 32 64 32 65 32 6F 32 70 ; EXCEPTION ROUTINE
3190 F3 EE FF 32 0D 13 6F 2B 29 2E A7 4A 35 2A 0C AC
```

PROGRAMMABLE MELODY MAKER

Mask-programmed ROM (Read Only Memory) ICs are capable of providing complex functions under low-power conditions. Unfortunately,

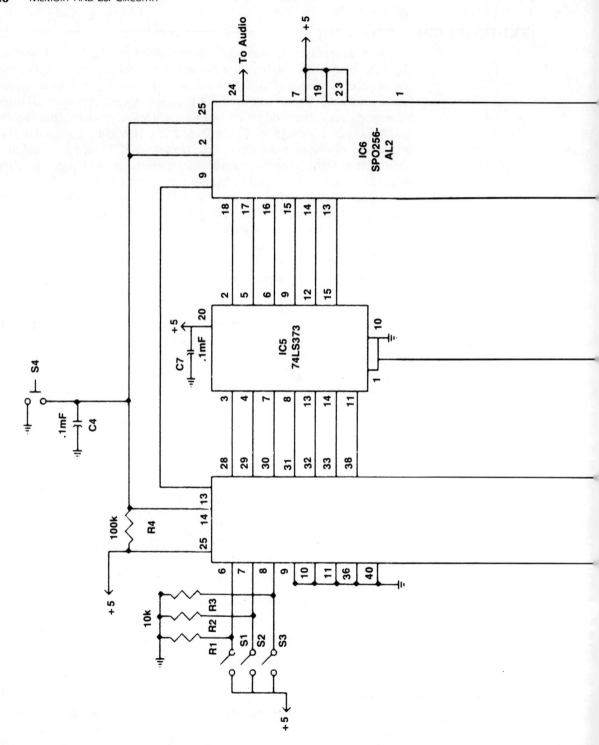

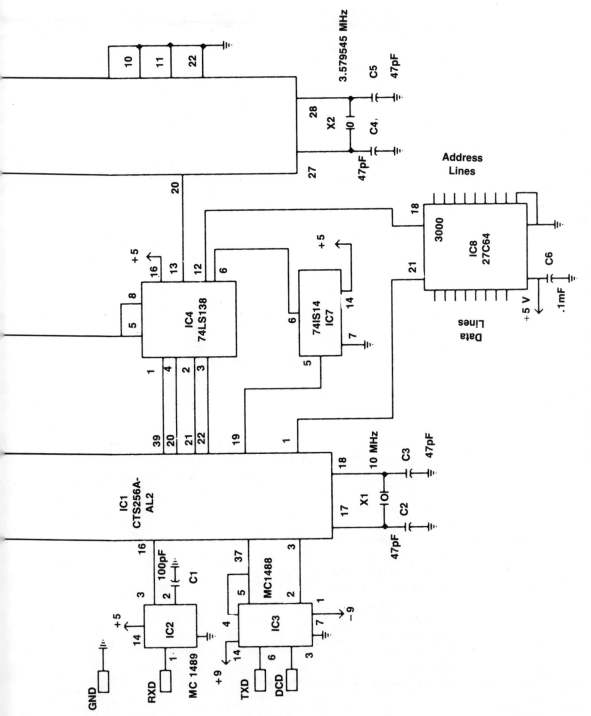

Fig. 10-10. Schematic diagram for Text-to-Speech Synthesizer.

these same ROMs are fixed in their output structure. In the case of the melody generator IC, however, the potential for a limitation from this fixed output can be reduced through the design of special application circuitry (see Fig. 10-11).

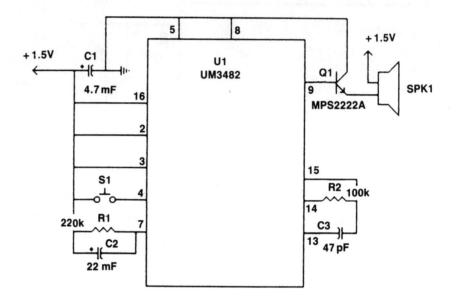

Fig. 10-11. Schematic diagram for Programmable Melody Maker.

This project is wired for the production of a different melody following each pressing of switch S1 (see Fig. 10-12 and Fig. 10-13). After S1 has been pressed, the song will play and automatically stop with its conclusion. There are a total of 12 different melodies that are possible with this PCB project:

"American Patrol"
"Rabbits"
"Oh, My Darling Clementine"
"Butterfly"
"London Bridge is Falling Down"
"Row, Row, Row Your Boat"
"Are You Sleeping"
"Happy Birthday"
"Joy Symphony"
"Home Sweet Home"
"Wiegenlied"
"Melody on Purple Bamboo"

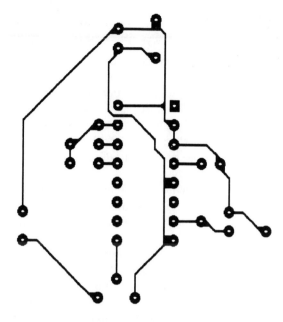

Fig. 10-12. Solder side of the Programmable Melody Maker PCB template.

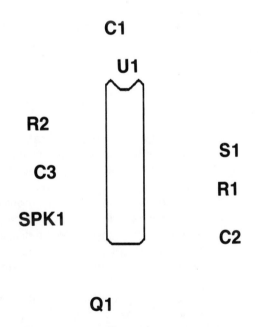

Fig. 10-13. Parts layout for the Programmable Melody Maker PCB.

EPROM PROGRAMMER ⎯⎯⎯⎯⎯⎯⎯⎯⎯⎯⎯⎯⎯⎯⎯⎯⎯⎯⎯⎯⎯⎯⎯⎯⎯⎯

Integrating an EPROM into a circuit design is more than just a matter of skillful schematic diagramming. An important consideration must also be given to the actual programming of the EPROM. Programming an EPROM can be most simply thought of as the writing or "burning in" of a predetermined set of binary data at a specified address location. Once this data/address combination has been established it will eventually become the fixed inner workings of an EPROM. Before any programming can take place, however, three important EPROM selection criteria must be fulfilled:

❖ Write and debug the program.
❖ Determine the final size of the program.
❖ Match an EPROM to the required program space.

Granted, these first two selection factors are obvious in their merit. Creating an EPROM-based program "on-the-fly" would be a ridiculous, if not, impossible proposition. Therefore, careful planning and evaluation of the final program must be made prior to committing this data to the floating gates of an EPROM. There are two courses of action for writing, debugging, and sizing an eventual EPROM program.

The first method is the most laborious. In this case, all of the program's data must be entered physically into the intended circuit. Analog switches, RAM, ROM, and/or microcomputer-based terminal control could be used for evaluating the validity of the programming. No matter which of these techniques that is used, however, only careful record keeping can hope to minimize the occurrence of error. Remember all of these test actions will need to be repeated during the programming of the target EPROM. As such, flicking analog switches can become extremely taxing after testing only a 64-byte program.

Another, far more practical, evaluation method utilizes the storage and processing abilities of a microcomputer. This method also offers an answer to the need for accurate and thorough record keeping. By using a microcomputer, the intended program can be repeatedly tested, altered, and corrected with the results saved onto a floppy disk medium. Later when the EPROM is ready for receiving the programming, the microcomputer can serve as the host for transmitting the data and address locations to the EPROM for their subsequent writing. This processing, storage, and transmission ability of the microcomputer can be an important ally when programming a 512K bit EPROM.

Following the solution to these first two steps in EPROM selection, the appropriate memory device must be chosen for receiving the programming. Faced with over 20 different EPROMs to choose from (this value is exclusive of the number of different EPROM manufacturers), the selection process can become cloudy. For the most part, however, the smaller memory organization size EPROMs can be eliminated. In other words, the 1702s, 2708s, 2716s, and 2732s offer such a limited amount of programming space that their use is impractical. A better solution would be found in the 2764s, 27128s, 27256s, and 27512s. These EPROMs contain the amount of data storage space that is capable of dealing with today's larger programming. For example, the 2764 is organized in an 8K × 8 or 64K bit memory structure. The 27128 is organized in a 16K × 8 or 128K bit memory structure. The 27256 is able to hold 32K × 8 or 256K bits of data. The 27512 can store up to 64K × 8 or 512K bits of data.

Another fringe benefit to using these larger capacity EPROMs is that many of them require a lower programming voltage. While this virtue isn't important when using pre-programmed EPROMs, its value does become readily apparent to the EPROM programmer.

EPROM PROGRAMMING

No matter which EPROM you select, there are a series of five simple programming steps that must be performed:

✤ Load the address on the EPROM's address pins.
✤ Load the program data on the EPROM's data pins.
✤ Disable the output enable.

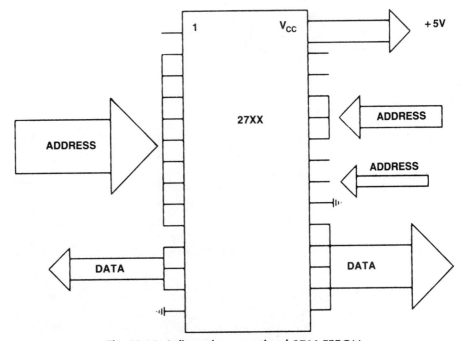

Fig. 10-14. A figurative operational 2764 EPROM.

❖ Apply the programming voltage.
❖ Pulse the program pin for approximately 50 mS.

Figure 10-14 is a representation of a typical EPROM. This figurative EPROM is pictured in its operational state. In other words, the power for this memory device is being drawn from the circuit with address and data control under the direction of an outside microprocessor.

During programming the nature of the EPROM's pins changes. Figure 10-15 shows the loading of the IC's address lines (A0 - A12). In its operational state, this EPROM's address lines are used for specifying a memory location for reading its stored data. In its programming state, however, these same address lines are used for specifying the memory location for *writing* the programmed data. A typical 2764 EPROM is capable of programming 8192 memory locations; locations 00h (0 0000 0000 0000) through 1FFFh (1 1111 1111 1111).

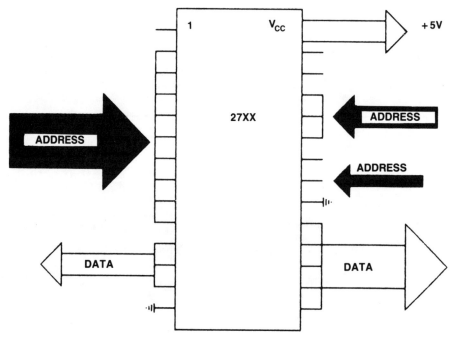

*Fig. 10-15. In step 1 of programming a
2764 EPROM, the address is loaded onto pins A0-A12.*

Once the address lines have been loaded with the correct memory location, the location's data must be placed on the data input lines (O0 - O7). In Fig. 10-16, the data output lines in the EPROM's operational state become the data input lines during programming. Within this illustration, O0 is the least significant bit (LSB) and O7 is the most significant bit (MSB).

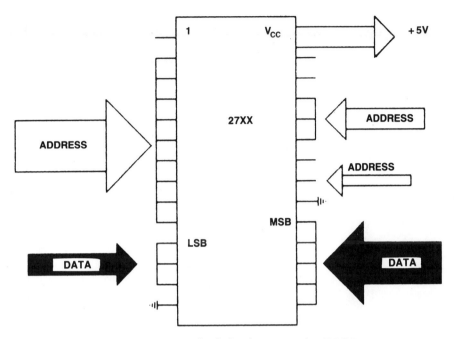

Fig. 10-16. Step 2, load the data onto pins D0-D7.

The final three steps in EPROM programming involve pin-level voltage changes. The first of these changes deals with both the output enable and chip enable pins. Figure 10-17 shows that these pins are connected to the system power supply and ground, respectively, during programming. This connection places the output enable in a TTL high state and the chip enable in a TTL low condition. In its operational state, the EPROM's output enable and chip enable are both low. Therefore, making the output enable pin high disables this function and allows data to be written to the EPROM.

After bringing the output enable pin high and chip enable pin low, the programming voltage is applied to programming voltage supply pin (see Fig. 10-18). For the most part, there are three different programming voltages that must be met for today's EPROMs: +25V, +21V, and +12.5V. Of these three power levels, only the +21V and +12.5V will be the most frequently used. These voltages cover the popular 2764, 2764A, 27128, and 27128A. Both of the "A" designated EPROMs (i.e., 2764A and 27128A) use the more convenient +12.5V programming voltage. Combining this modest programming voltage requirement with its large memory capacity, the 27128A should become your EPROM of choice.

The final step in the programming of an EPROM deals with a close tolerance pulse applied to the programming pin (see Fig. 10-19). This pulse, a high-to-low-to-high fluctuation, is the programming circuit's +5V TTL power

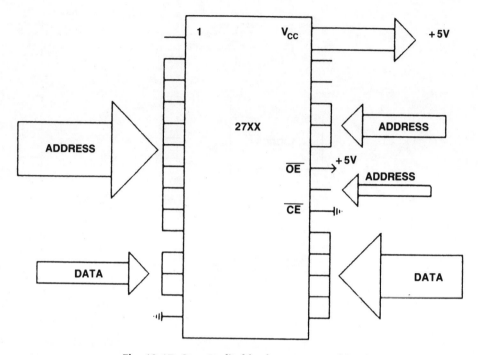

Fig. 10-17. Step 3, disable the output enable pin.

of this pulse is critical to the successful programming of the EPROM. A standard of 50 mS (with a minimum of 45 mS and 55 mS being an absolute maximum) pulse duration is found on 2764, 2764A, 27128, and 27128A EPROMs. The accurate application of this pulse results in the latching and burning in of the set data at the specified address location.

One problem with this pulse programming method is the length of time that is required for programming an EPROM. For example, based on the 50 mS pulse duration, a 27C128 EPROM requires approximately 13.7 minutes for complete programming (this time value is exclusive of the time that is needed for setting up the address and data lines). A solution to this time factor is available from several manufacturers. By using a fast programming algorithm based on a closed-loop verification cycle, EPROM programming times can be reduced dramatically. In operation, this fast algorithm uses an initial 1 mS pulse which is followed by a verification check. This initial pulse is repeated until the verification is positive. Once the check is correct, an over-program pulse of 3 to 4 times the initial pulse number is applied. In other words, if eight 1 mS initial pulses were required for proper testing, then a 24 to 32 mS over-program pulse would be applied. Therefore, a complete programming cycle with an average number of tests would take approximately 9.3 minutes on a 27C128 EPROM.

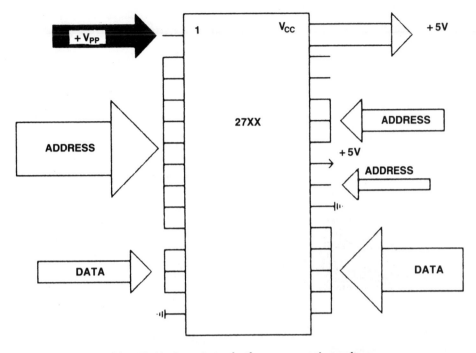

Fig. 10-18. Step 4, apply the programming voltage.

A final requirement for fast-programming EPROMs is the manditory change in the IC's supply voltage. In normal programming and operation, this voltage is +5V. During fast programming, however, this voltage must be increased to +6V. Following the entire fast programming sequence of initial and over-program pulsing, the supply voltage must be returned to the native +5V state.

CONSTRUCTION OF THE EPROM PROGRAMMER

An extremely simple EPROM programmer can be constructed from a limited number of parts. The EPROM programmer is a low-cost programmer that is able to write, verify, and read 27C64 and 27C128 EPROMs. As a cost-cutting measure, each of these operations is performed by hand through analog switches and LEDs. While this method does greatly reduce programming speed, the EPROM programmer is more than adequate enough for building all of the projects that are discussed in this book.

Only six ICs are necessary for building the EPROM programmer (see Fig. 10-20). Central among these chips is the 555 Timer. This IC, when coupled with two capacitors and a resistor, generates the 50 mS program pulse. This generated pulse is an inverse low-high-low pulse, therefore, the signal is routed through a 7406 Hex Inverter Buffer/Driver with open-collector high-voltage output. This IC converts the pulse into the needed high-low-high pulse.

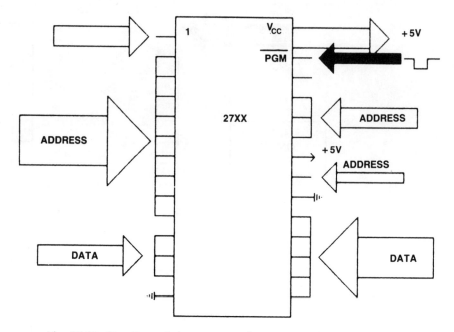

Fig. 10-19. Step 5, send the programming 50 mS pulse to the EPROM.

Begin construction by soldering five IC sockets to the selected board. Next the ZIF (Zero Insertion Force) socket should be added. Follow the attachment of these sockets with the remaining resistor packs, capacitors, switches, LEDs, and potentiometer. Before you press the ICs into their sockets, make all of the power connections to the EPROM programmer.

TESTING THE EPROM PROGRAMMER

There are two levels of testing before this project can become a bona fide EPROM programmer. First, all of the power connections should be verified for their indicated voltages. Use the following procedure for testing the EPROM programmer voltages:

1. Make the power connections.
2. Switch the power switch on. NOTE: Immediately turn this switch off, if you notice any excessive heat or smoke.
3. Obtain a Multimeter.
4. Compare each voltage with the correct results.

If each of these voltage tests has been performed correctly, then all of the EPROM programmer ICs can now be inserted into their respective sockets. Now, with all of the ICs in their sockets, each of the voltage tests from steps

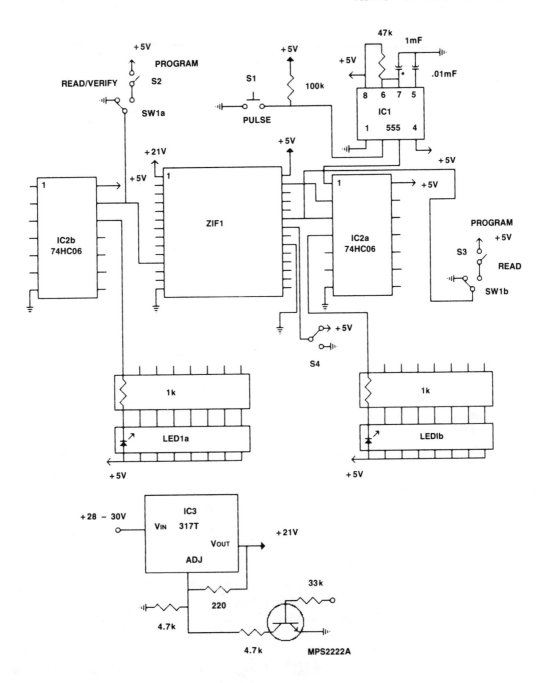

NOTE: Repeat I/O connections for all address and data lines.

Fig. 10-20. Schematic diagram for the EPROM programmer.

1-4 should be repeated. This time, however, use a logic probe for verifying the low and high status of each test location.

The final test for the completed EPROM programmer is an operational test. This test can be performed with either a 27C64 or a 27C128 EPROM inserted in the ZIF. Remember, if you are testing the 14 address line 27C128, to make an amendment to your test procedure that will accommodate the extra address line. Use the following procedure for testing the programming, verifying, and reading states of the EPROM programmer:

1. Use the switch settings and LED readings in Fig. 10-21 for conducting this battery of tests.
2. Read the address locations listed in Table 10-1.
3. Program each address location with its associated data as listed in Table 10-2.
4. Verify each programmed address location by reading the data from each memory location.

A careful construction technique will yield a 100 percent operational EPROM programmer. Should any of the above tests fail to execute properly, however,the culprit might be either the programming voltage or the programming pulse. A quick test from a multimeter will answer the programming voltage question. Unfortunately, two problems can plague the programming pulse. The first is the pulse shape and the second fault might be the pulse duration. In either case, an oscilloscope test will be the only method for accurately determining the true shape and duration of this pulse.

USING THE EPROM PROGRAMMER

There are three operations that can be performed with the EPROM programmer: read, program, and verify. In this application, the read and verify operations are performed in the same manner.

Read

1. Place the EPROM programmer in the program mode (PROGRAM).
2. Neutralize the data switches.
3. Load the desired address location on the address switches.
4. Read the LED for the binary representation of the data at the loaded address location.

Program

1. Place the EPROM programmer in the read mode (READ/VERIFY).
2. Set the data switches to the correct binary representation.
3. Load the desired address location on the address switches.
4. Press the program button (PULSE).

Table 10-1. READ Address Locations for Testing the EPROM Programmer.

EPROM Programmer Read Address Locations	
Address	**Data**
00000000	
00001111	
11110000	
10101010	
01010101	
11111111	
1000000000000	
1000000000001	
1111111111111	
*For the 27128:	
10000000000000	
10000000000001	
11111111111111	

Table 10-2. PROGRAM Address/Data Locations for Testing the EPROM Programmer.

EPROM Programmer Program Address Locations	
Address	**Data**
00000000	00000000
00001111	00110011
11110000	11001100
10101010	00110011
01010101	11001100
11111111	00000000
1000000000000	11001100
1000000000001	00110011
1111111111111	00000000
*For the 27128:	
10000000000000	11001100
10000000000001	00110011
11111111111111	00000000

DATA

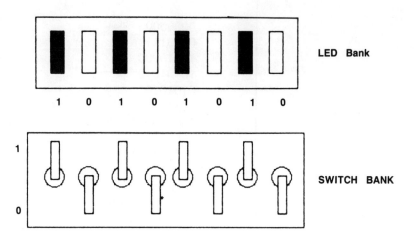

ADDRESS

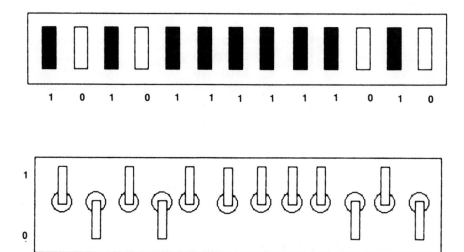

Fig. 10-21. LED readings for testing the EPROM Programmer.

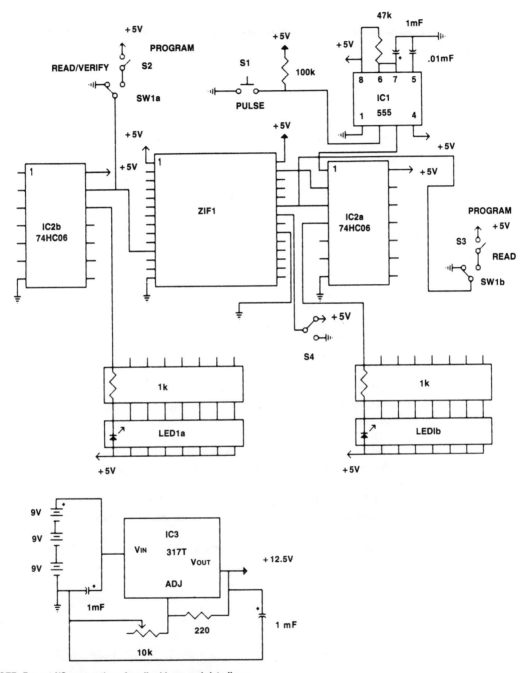

Fig. 10-22. Schematic diagram for the EPROM Programmer II power supply.

NOTE: Repeat I/O connections for all address and data lines.

Verify

1. Place the EPROM programmer in the verify mode (READ/VERIFY).
2. Neutralize the data switches.
3. Load the desired address location on the address switches.
4. Read the LED for the binary representation of the data at the loaded address location.

THE EPROM PROGRAMMER II

Whereas the EPROM programmer was only able to run from household line current and program 27C64 and 27C128 EPROMs, the EPROM programmer II is able to operate with battery power and program 27C256 and 27C512 EPROMs. The major advantage of using these suffix higher memory capacity EPROMs is the reduced programming voltage. A modest +12.5V on pin #1 is enough voltage for programming these memory devices. Furthermore, a nice power supply that is able to provide both the circuit voltage, as well as the +12.5V programming voltage can be constructed from a battery source.

All of the operational specifications for the EPROM programmer II are identical to those found on the original EPROM programmer. Only a minor modification to the power section of this EPROM programmer circuit is required (see Fig. 10-22).

Operationally, the EPROM programmer II follows the same read, program, and verify procedures as described for the EPROM programmer. There are only two important functional differences that are inherent to the EPROM programmer II, however. First, the battery power supply must be kept fresh and within a 1 - 2 volt tolerance level. Failure to maintain the stated supply and programming voltage levels will lead to EPROM writing errors. Testing the voltages with a multimeter prior to use is an important safeguard against this problem.

The second point to consider during the operation of the EPROM programmer II is that this programmer has been designed for reading and writing 27C256 and 27C512 EPROMs, only. Trying to program any other EPROM type can result in damage to both the EPROM and the EPROM programmer II.

11

CMOS Circuitry in the Microprocessor Environment

CDP1802 CMOS EIGHT-BIT MICROPROCESSOR

The CDP1802 CMOS Eight-Bit Microprocessor is an 8-bit register LSI (Large-Scale Integration) CPU (Central Processing Unit) which is capable of addressing a maximum combined RAM/ROM memory map of 65,536 bytes. The 8-bit parallel data port is a bidirectional bus featuring a multiplexed address bus. A 16 × 16 register map provides program counters, data registers, and data flags. Memory access and control is monitored through an internal DMA (Direct Memory Access), a memory interrupt, and a flag input. There is a 91 member mnemonic CPU instruction set for controlling the actions of the CDP1802.

Package Configuration: 40-lead DIP

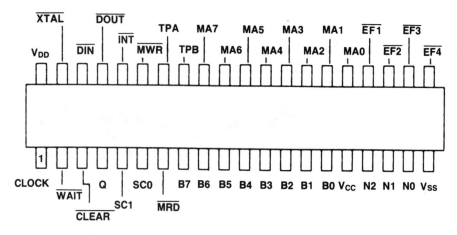

Fig. 11-1. Pin assignments for CDP1802 CMOS 8-Bit MPU.

CPU Registers: 8-bit Data Register (D)
1-bit Data Flag (DF)
4-bit Program Counter (P)
4-bit Data Pointer (X)
8-bit Holding Register (B)
16-bit Scratchpad (R)

Clock Frequency: 3.2 MHz

Output Capacitance: 10 pF

Data Retention Current: 0.05 mA

Clock Pulse Width: 60 - 150 ns

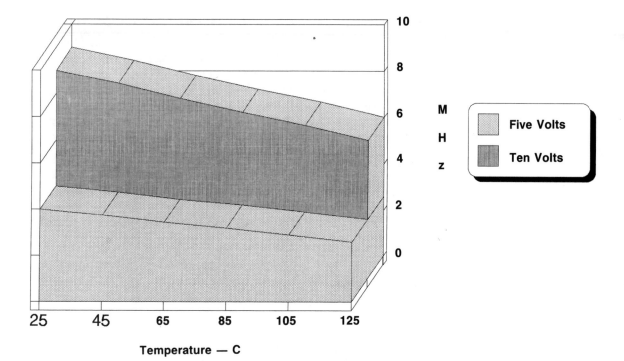

Temperature — C

Fig. 11-2. Clock frequency for CDP1802.

Clear Pulse Width: 75 - 300 ns

Data Setup Time: 25 - 40 ns

Data Hold Time: 75 - 200 ns

Propagation Delay Time; Clock-to-Output: 100 - 400 ns

80C39 CMOS SINGLE-CHIP EIGHT-BIT MICROCOMPUTER

The 80C39 CMOS Single-Chip Eight-Bit Microcomputer is an 8-bit register MPU (Microprocessing Unit) capable of addressing a maximum combined RAM/ROM memory map of 65,536 bytes. Dual 8-bit parallel data ports and one data bus can be accessed through 12-bit addressing. An eight level stack, featuring 8 pairs of registers, provides program counters, data registers, and data flags. Memory access and control is monitored through an internal 8-bit interval timer and event counter. There is a 98 member mnemonic CPU instruction set for controlling the actions of the 80C39.

Package Configuration: 40-lead DIP

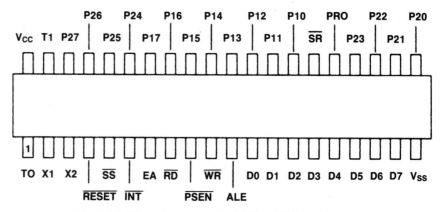

Fig. 11-3. Pin assignments for 80C39 CMOS 8-Bit MPU.

Read/Write Pulse Width: 480 ns

Data Setup Time: 390 ns

Data Hold Time: 40 ns

Data Delay Time: 350 ns

Address Setup Time to ALE: 70 ns

Address Setup Time from ALE: 50 ns

Cycle Time: 1.36 ms

80C85 CMOS EIGHT-BIT MICROPROCESSOR

The 80C85 CMOS Eight-Bit Microprocessor is an 8-bit register CPU that is capable of addressing a maximum combined RAM/ROM memory map of 65,536 bytes. The 8-bit parallel data port is a bidirectional bus featuring a multiplexed data bus. Twelve addressable 8-bit registers provide two pairs of 16-bit registers and six interchangeable 8- or 16-bit registers. The CPU instruction set is compatible with the instruction set of the 8080A CPU.

Package Configuration: 40-lead DIP

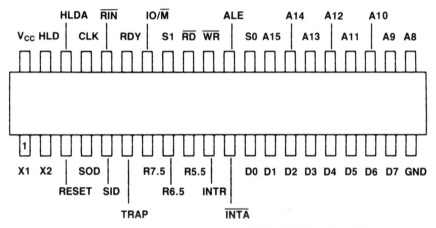

Fig. 11-4. Pin assignments for 80C85 CMOS 8-Bit MPU.

CPU Registers: 8-bit Accumulator (ACC)
Five Flags (F)
16-bit Program Counter (PC)
16-bit X 3 Data Pointer (HL)
16-bit Stack Pointer (SP)

Clock Reference: 32.768 kHz

80C86 CMOS SIXTEEN-BIT MICROPROCESSOR

The 80C86 CMOS Sixteen-Bit Microprocessor is a 16-bit register MPU capable of addressing a maximum combined RAM/ROM memory map of 1M bytes. The 16-bit parallel data port is a bidirectional bus with a 14-word-by-16-bit register set featuring symmetrical operations. Both signed and unsigned arithmetic operations are possible in 8- or 16-bit, and binary or decimal modes. There is a 90 member mnemonic MPU instruction set that is compatible with the 8086 MPU for controlling the actions of the 80C86.

Package Configuration: 40-lead DIP

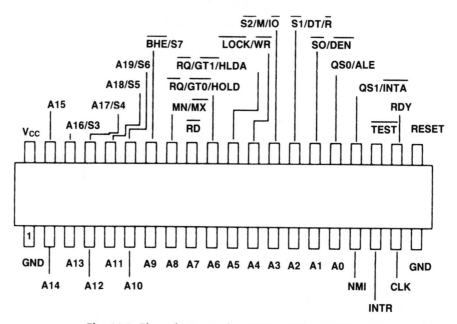

Fig. 11-5. Pin assignments for 80C86 CMOS 16-Bit MPU.

Clock Frequency: 5 MHz

Output Capacitance: 15 pF

MPU Registers: 8-bit Accumulator (AL)
16-bit Accumulator (AX)
16-bit Count (CX)
8-bit Count (CL)
16-bit Data (DX)
8-bit Data (DL)
16-bit Base (BX)
8-bit Base (BL)
16-bit Stack Pointer (SP)
16-bit Base Pointer (BP)
16-bit Source Index (SI)
16-bit Destination Index (DI)

Clock Rise/Fall Time: 10 ns

Data Setup Time: 30 ns

Data Hold Time: 10 ns

Address Hold Time: 10 ns

80C88 CMOS EIGHT-BIT MICROPROCESSOR

The 80C88 CMOS Eight-Bit Microprocessor is an 8-bit data/16-bit internal register MPU capable of addressing a maximum combined RAM/ROM memory map of 1M bytes. The 8-bit parallel data port is a bidirectional bus with a 14-word-by-16-bit register set, featuring symmetrical operations. Both signed and unsigned arithmetic operations are possible in 8- or 16-bit, and binary or decimal modes. There is a 90 member mnemonic MPU instruction set that is compatible with the 8088 MPU for controlling the actions of the 80C88.

Package Configuration: 40-lead DIP

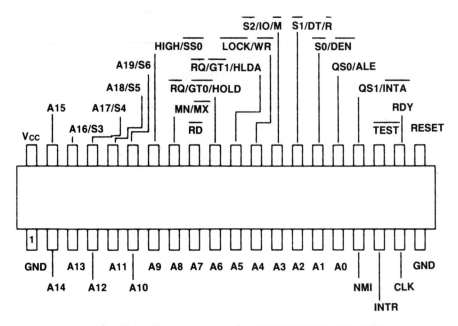

Fig. 11-6. Pin assignments for 80C88 CMOS 8-Bit MPU.

Clock Frequency: 5 MHz

Output Capacitance: 15 pF

MPU Registers: 8-bit Accumulator (AL)
16-bit Accumulator (AX)
16-bit Count (CX)
8-bit Count (CL)
16-bit Data (DX)
8-bit Data (DL)
16-bit Base (BX)
8-bit Base (BL)
16-bit Stack Pointer (SP)
16-bit Base Pointer (BP)
16-bit Source Index (SI)
16-bit Destination Index (DI)

Clock Rise/Fall Time: 10 ns

Data Setup Time: 30 ns

Data Hold Time: 10 ns

Address Hold Time: 10 ns

HEATH'S ET-3400A TRAINER

No matter which breed of microcomputer you own, there are several underlying similarities that make them all birds of the same feather. Basically, every computer, including Apples and IBMs, shares three general characteristics. First, they all have the same elemental chip architecture. This is not to imply that all microcomputers use identical components. For example, the Apple Computer Macintosh uses a Motorola 68000 MPU (microprocessor unit), while the IBM PC AT relies on the Intel 80286 MPU. In point of fact, this dissimilarity actually illustrates the similarity found in all microcomputers—they all employ an MPU. Likewise, there are RAM (random access memory), ROM (read-only memory), and I/O (input/output) interface sections each populated with different IC (integrated circuit) nomenclature, but sharing an identical architectural function.

The next area of joint microcomputer similarity is at the binary logic level. Even though this shared attribute sounds too simplistic to be of any practical value, understanding the interrelationship between the internal movement of 1's and 0's, and the external expression of this action simplifies a user's comprehension of a computer's operation.

Finally, the third characteristic found on all microcomputers is the ability to expressively manipulate each IC architectural section with the high-level language interface that is available through hexadecimal programming. This base 16 numbering system presents the user with yet another method of controlling the actions of the computer.

Unfortunately, there is a problem with trying to deal with these three areas at the commercial microcomputer level. Basically, this problem centers around an issue of complexity.

Most computer manufacturers have buried the computer's architecture, hidden its binary logic, and insulated the user from hexadecimal programming with the intent of making their product less complex, and hopefully, ensure its commercial viability. Therefore, if comprehending these three levels of microcomputer operation piques your interest, you will need to go elsewhere for your education.

Luckily, you don't need to look very far for answering these three basic computer questions. As is becoming more frequently the case, the Heath Company of Benton Harbor, Michigan is the prime mover in supplying this type of pedagogical computer material. In this case, the Heath Model ET-3400A Microprocessor Trainer is a fully-functional 8-bit computer with one big difference. Namely, it is a completely open system—both figuratively as well as literally. This design makes it a snap for realizing a thorough introduction to the function and operation of a microcomputer on its architectural, binary, and hexadecimal levels. Furthermore, using Heath's Microprocessor Trainer provides you with the opportunity to attempt microprocessor-related experiments that might otherwise prove detrimental to the well-being of your main microcomputer system. Along these lines Heath supplies five different experimenter's courses that interact with the completed Trainer on a purely "nuts 'n bolts" level. Therefore, a complete computer education can be had through the assembly and use of the Microprocessor Trainer.

INSIDE THE ET-3400A

There are few secrets in the design or structure of the Microprocessor Trainer. All of the major computer sections (e.g.; RAM, ROM, and I/O) are physically located on the top panel of the Trainer. This "open-air" construction technique eliminates the need for constantly removing a chassis top for accessing the various interface sections during experimentation.

The Microprocessor Trainer is nothing short of a complete microcomputer in its own right. Beginning with its control circuitry, where the MPU for directing the Trainer's operations, the 8-bit Motorola MC6808 (a Hitachi HD6802 variant was the MPU supplied with this review kit) running at a 1 MHz clock frequency can be found. This IC is an upgrade from the original Trainer's (Model ET-3400) MC6800 MPU. Essentially, the difference between these two MPUs amounts to the presence of an internal clock oscillator/driver on the MC6808 (or, HD6802, as is the case with this particular kit). Other than this hardware difference, both MPUs remain software compatible, and use the same instruction set.

Supporting the MC6808 are 512 bytes of NMOS (N-Channel Metal Oxide Semiconductor) RAM. The two ICs (actually two 1024 × 4-bit 2114 ICs) that constitute this section are located in a pair of top panel sockets permitting easy expansion of the Trainer's usable memory.

A 1K byte 2716 EPROM (this is actually a 2K × 8-bit programmable ROM circuit with only one-half of its available memory occupied by programming) dominates the Trainer's ROM section. This IC houses a monitor program that coordinates the hexadecimal programming of the Trainer. As a considerate gesture, Heath Company features the complete listing for this monitor program within the Trainer's assembly manual. By studying this monitor program listing, alternate EPROMs can be burned in and thereby supply different attributes to the Trainer's 17-key keypad. As a word of caution, in order for these customized EPROMs to function properly, they must be 24-pin-compatible with a 2716 EPROM.

Several I/O sections are available for gaining access to the MC6808, EPROM, and RAM circuits. Foremost among the input interfaces is the 17-key keypad. Fifteen of the seventeen (1-F) keys perform two separate functions with their primary assignment being numerical entry points. Their other function is derived from the monitor program that is contained within the 2716 EPROM. These assignments include display accumulator contents (keys 1 and 2), set breakpoints (key 9), and execute a program (key D). The two remaining keys on the keypad, the 0 key and the RESET key, have single function assignments (0 enters the numeric value 0 and RESET halts program entry and program execution).

One other switchable input device is an 8-switch DIP (dual in-line package) that is found in the binary data section of the Trainer. This switch bank is used for controlling the flow of binary logic to a set of terminal output blocks and light LEDs (light emitting diodes). Each switch from the 8-position DIP is assigned to a single LED. These eight LEDs are then individually buffered for selective I/O logic monitoring.

In order to monitor all of the various input functions, a bank of six 7-segment LED digits provide a constant readout of the Trainer's various registers, accumulators, and addresses. Each of these LED digits is socketed on the Trainer's main circuit board. The removable nature of these LED digits provides a controlling output from various MPU functions and hexadecimal programming. In fact, this control ability is stressed in many of the experiments that accompany the five Heath Microprocessor Trainer related educational courses.

One final laudable feature of the Microprocessor Trainer is its "open" system architecture. By placing special multi-position connector blocks at

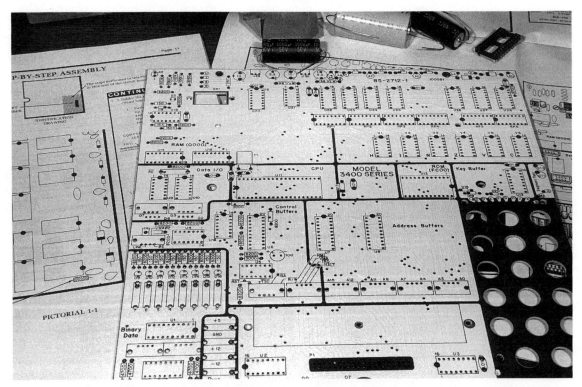

Fig. 11-7. Assembling Heath's ET-3400A Trainer is ably supported by superb documentation and expertly silkscreened PCBs.

Fig. 11-8. There are 58 resistors and six resistor packs
which must be soldered to the main PCB of the Trainer.

identified input, output, data, and address lines, the Trainer user is able to delve into the inner workings of an MPU without subjecting the circuitry to endless hardware modifications. For example, one connector block in the control buffer section of the Trainer provides outputs for the $\phi2$ and VMA$\phi2$ clocks as well as a 1Hz square wave. Therefore, timing experiments can be run directly from the MPU or controlled with a fixed wave source.

ASSEMBLY OF THE ET-3400A

Simplicity is an important virtue found in the construction phase of all Heath kits. Even though there are several hundred parts, two circuit boards, and 28 feet of connecting wire the lucid, and heavily illustrated Heathkit Manual and Illustration Booklet expertly guide the user through every construction detail (see Fig. 11-7).

Heath has divided the construction of the Trainer into four distinct assembly subsections: the main circuit board, the keyboard circuit board, the support bracket, and the cabinet assembly and wiring. A separate construction chapter is used for the assembly instruction for each of these subsections.

During the construction, the main circuit board serves as the principal site for attaching each of the subsections. The following step-by-step assembly description outlines Heath's logic in employing this construction technique.

1. All support components and IC sockets are soldered into position on the main circuit board (see Fig. 11-8).
2. Group soldering techniques are used for attaching several components at the same time. For example, six IC sockets are inserted into the component side of the main circuit board. Following this insertion process, the board is flipped over and all six of the sockets are soldered into place at the same time.
3. Resistors and diodes are soldered into place, first. Later, the IC sockets are added.
4. The final additions to the main circuit board are the jumper blocks, switches, and LEDs (see Fig. 11-9).
5. The keyboard circuit board is a separate assembly that holds the hexadecimal keypad's keys (see Fig. 11-10).
6. The completed keyboard circuit board is fixed to the main circuit board with four screws.

Fig. 11-9. Following the connection of the support components several jumper blocks are soldered onto the main PCB.

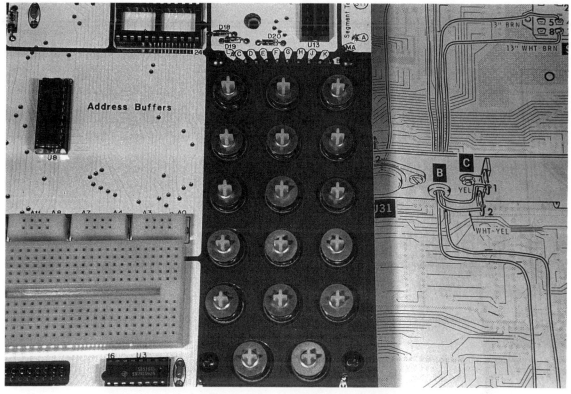

*Fig. 11-10. The completed keypad is connected to the bottom of the PCB through
12 address lines and fixed in place with four screws.*

7. A preformed piece of aluminum serves as the power supply's support
   bracket.
8. Three electrolytic capacitors and a voltage regulator are held in place
   on the two sides of the support bracket (see Fig. 11-11).
9. The final cabinet assembly and wiring begins with the construction
   of the power supply box, and the installation of the power
   transformer.
10. Finally, the power supply lines are soldered onto the main circuit
    board and the ICs are seated in their sockets.

Only three minor errors were noted during the construction of the
Microprocessor Trainer. Of the three, only one of these errors was hardware
oriented, and could cause a problem during the later testing stages and final
operation of the Trainer.

Error 1. IC U1 is not labeled in Pictorial 1-13.

*Fig. 11-11. The main power supply capacitors
are held in place with an aluminum support bracket.*

Error 2. One of the post holes for receiving the 8-pin connector block J19 is too small. This problem should be corrected by shaving a small amount of plastic from J19's post.

Error 3. Pictorial 4-3 displays an incorrect bolt/nut orientation at Point C and fails to show the 2 electrolytic capacitors on the support bracket.

TESTING YOUR ASSEMBLY

Once all of the soldering has been completed, a series of three initial tests are performed on the assembled Trainer. The first test is a voltage test that confirms the correct construction of the Microprocessor Trainer's power supply. A volt-ohmmeter is used as the examination device during this test phase. If you don't have access to a volt-ohmmeter, this series of tests can be ignored without jeopardizing the integrity of the completed Trainer.

The next battery of tests verify the operation of the LED indicators. Following the evaluation of the six main LED digits, the eight LEDs found in the binary data section are examined by selectively lighting and extinguishing each LED. Prior to the testing of these eight LEDs, the RAM,

ROM, MPU, and key buffer ICs are each placed in their respective sockets. After these memory circuits have been added, the Binary Data LEDs are examined. This test also checks the soldering connections between the 8-position DIP, the Binary Data jumper blocks, and the LEDs.

If no errors occur during both of these tests, then the final operational tests are performed on the Trainer. This last group of exercises requires the keypad input of three short hexadecimal programs that report the status of the Trainer keypad and internal memory circuits. For example, the first 11-step program verifies that all seventeen keys on the keypad are operating properly. If this program executes correctly, a 15-step program checks the Trainer's memory circuits by manipulating Accumulator A. Finally, the third program in this operational test makes the top segment of the farthest left LED glow when it is executed (this is performed through Heath's DO command).

After all three of these tests have been performed, and if the results are all positive, the Microprocessor Trainer's final assembly is undertaken. Basi-

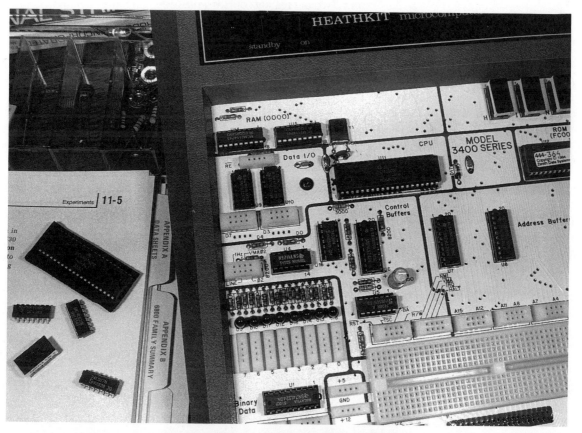

Fig. 11-12. Several packaged Heath experiments can be performed on the completed Trainer.

cally, this is just a matter of mating the top and bottom cabinets together, and affixing several labels to the completed chassis. Finally, when the FCC label has been properly attached to the bottom of the chassis, the Heath Microprocessor Trainer is ready for operation.

The completed Trainer was used in a normal educational capacity for a period of three weeks. During this review period, only one unexpected, if not disagreeable, phenomenon was witnessed. In the first three hours of use, a distasteful odor was emitted from inside the Trainer's chassis. After a lengthy examination of the power supply, it was discovered that some paint and plastic in the vicinity of the power transformer was being vaporized and, therefore, was producing the foul smell. Interestingly enough, after the three hour "burn-in" period, the odor disappeared and the Trainer has operated with nary a bad smell ever since.

A fitting testament to the value of Heath's Microprocessor Trainer can be found in the five educational courses that are specifically designed for use with the Trainer. Two of Heath's courses were tested on the assembled Trainer: Microprocessor Interfacing (see Fig. 11-12) and Voice Synthesis. All of the experiments supplied with both of these courses executed flawlessly. As you have come to expect from a Heathkit product, each course is accompanied by a complete set of parts, thorough documentation, and, where it is applicable, detailed software. For example, with the Voice Synthesis course, the Microprocessor Trainer was able to utter its first spoken words in less than one hour by following the instructions found in Experiment 1. This is indeed a remarkable accomplishment for such a primitive device.

A

Building a CMOS Project

Building your own electronic circuit is an exciting project that has been carefully detailed in this book's preceeding chapters. By combining a few dollars worth of parts with a circuit's schematic diagram, an intelligent design is born. Unfortunately, several factors are bound to block the successful implementation of this project. There are financial, mental, and physical reasons that might prevent the introduction of PCBs. While I can't satisfy either your financial or physical difficulties, I can try to lessen the severity of your naiveté to electronics construction techniques.

Whether you are a seasoned electronic project builder, or a casual computer user, you will need to learn a few construction basics. Most of these building techniques center around the most effective means for translating a circuit from paper into a wired, operating unit. Two specific areas in which this construction education will be stressed are the manner in which the circuit is built and the enclosure in which it is placed.

Seasoned electronics project builders may be experienced hands at soldering and printed circuit board (PCB) etching (if you aren't, read further in this appendix), but use of the E-Z Circuit boards is one method of project wiring that is available to all levels of project builders. Circuits are easily transferred from the printed schematic onto an E-Z Circuit board by using easily cut dry-transfer shapes and pad soldering. Circuit builders wishing more stability for their projects can quickly translate the E-Z Circuit designs onto an etched PCB.

Once you have finalized your design, it is time to place your circuit inside a housing. As a rule, a housing will only be necessary for stand-alone and parallel or serial port connection computer-based projects. For example, if your final project is for an internal expansion slot of a personal microcomputer, then you will not need to worry about a housing. Conversely, if your circuit uses a parallel port for interfacing with microcomputers, then you will need to consider the design of a circuit housing. In some ways, the familiar, boxy metal or plastic electronics project cabinet has turned into an antiquated relic. For the majority of the constructed projects, however, the purchased storage cabinet is the ideal project housing solution.

IF IT'S EASY, THEN IT MUST BE E-Z

Custom circuit boards used to be the exclusive domain of the acid etched PCB. Now Bishop Graphics has introduced a revolutionary concept that could become the next dominant circuit board construction technique. E-Z Circuit, that's what Bishop Graphics calls it, takes a standard, pre-drilled universal PC board and lets the builder determined all of the tracing and pad placements. A special adhesive copper tape is the secret to this easy miracle in PCB fabrication.

The E-Z Circuit system is both an extensive set of these special copper tape patterns and several blank universal PC boards (see Fig. A-1). The blank boards serve as the mounting medium for receiving the copper patterns. There

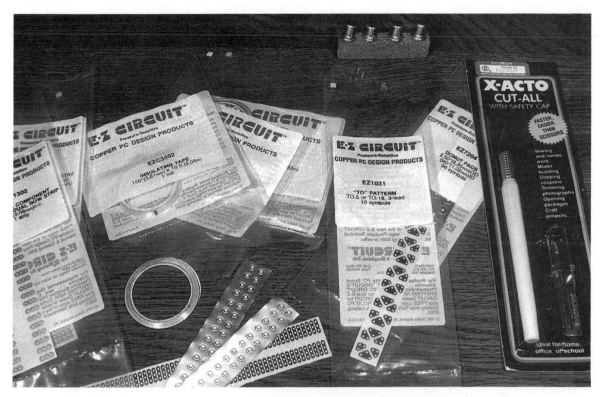

*Fig. A-1. Bishop Graphics features a thorough line of
E-Z Circuit adhesive copper patterns for quick and easy PCB fabrication.*

are general purpose blank boards (Bishop Graphics' #EZ7402 and #EZ7475) and Apple II family blank PC boards (Bishop Graphics' #EZ7464). Each of these boards is made from a high-quality glass epoxy and pre-drilled with IC spaced holes. There is also a blank edge connector region on each board for attaching one of E-Z Circuit's copper edge connector strips.

Complementing the blank boards is a complete set of copper patterns. Each pattern is supplied with a special adhesive that permits minor repositioning, but holds firmly once its position is determined. This adhesive also has heat resistant properties that enable direct soldering contact with the pattern. Only an extensive selection of patterns and sizes would make an E-Z Circuit design worthy of consideration for project construction. Once again, E-Z Circuit satisfies all of these requirements with edge connector, IC package, terminal, test point strip, tracing, donut pad, elbow, TO-5, power and ground strip, and power transistor patterns. Additionally, each of these patterns is available in several sizes, shapes, and diameters.

Only four simple steps are needed in the construction of an E-Z Circuit project. In step one, you will determine which pattern you need and prepare it for positioning. Step two is for placing the selected pattern on the blank PC board. This is a simple process that involves the removal of a flexible release layer from the back of the copper-clad pattern. During step three you insert all of the components into their respective holes. Finally, in step four, you solder the component's leads to the copper pattern. This step should be treated just like soldering a coventional copper pad or tracing. If you use a reasonable soldering iron temperature, anywhere from 400 to 600 degrees Fahrenheit, you won't need to worry about destroying the adhesive layer and ruining the copper pattern.

Making a slight digression on the issue of soldering irons, be sure to use an iron with a rating of 15 to 25 watts. Soldering irons that are matched to the demands of working with E-Z Circuit include: ISOTIP 7800, 7700 (the specifications for these two irons "border" the stated E-Z Circuit requirements, but they will work), and 7240, UNGAR SYSTEM 9000, and WELLER EC2000.

The result of this four step process is a completed project in less time than a comparably prepared acid etched PCB. Incidentally, the cost of preparing one project with E-Z Circuit versus the etched route is far less. Of course, this cost difference is skewed in the other direction when you need to make more than one PCB. This is because E-Z Circuit is geared for one-shot production and not assembly-line production. The bottom line is, before you decide on your circuit board construction technique (etched PCB or E-Z Circuit), read the remainder of this appendix for the latest advances in acid etched PCB design. Then evaluate your needs and resources and get to work on your first powerful project.

TAMING THE SOLDER RIVER

Before the first bit of solder is liquefied on your PC board, an overall concept must be organized for fitting the completed project into an enclosure. Small enclosures are usually preferred over the larger and bulkier cabinets simply because of their low profile and discreet appearance. Space is limited within the narrow confines of such a sloped enclosure, however.

It's quite easy to envision a two dimensional circuit schematic diagram and then forget that the finished, hard-wired circuit will actually occupy three dimensions. Capacitors, resistors, and ICs all give a considerable amount of depth to a finished project PC board. Fortunately, the effects of these tall components can be minimized through some clever assembly techniques.

With only a few exceptions, all components must be soldered to a PC board as closely as possible. One exception to this rule is in leaving adequate jumper wire lengths for external enclosure mounted components, such as switches and speakers.

Both component selection and their mounting methods directly affect a project's PC board depth. For example, the selection of a horizontally-oriented, miniature PC mountable potentiometer over a standard vertically oriented potentiometer may save up to ½ inch off of a board's final height. Likewise, flat, rectangular metal film capacitors offer a space savings whenever capacitors of their value are required (usually .01 μF to 1.0 μF).

If a disc or monolithic capacitor is used on a circuit board, the lead can be bent so that the capacitor lies nearly flat against the PC board. The capacitor can be pre-fitted before soldering it into place and the required bends can be made with a pair of needle-nosed pliers. Be sure that the leads of every component are slipped through the PC board's holes as far as possible before soldering them into place. Excessive leads can be clipped from the back side of the board *after* the solder connection has been made. At this point a precautionary note pertaining to overly zealous board compacting is necessary. Do not condense a board's components so tightly that undesired leads might touch; a short circuit is the inevitable and unwanted result for this carelessness. Also, some components emit heat during operation. Therefore, some component spacing is mandatory for proper ventilation.

One way to minimize PC board component crowding is by using a special mounting technique with resistors and diodes. The common practice for mounting resistors and diodes is to lay them flat against the PC board. This technique is impractical, and occasionally impossible, on the previously described PC boards. A superior technique is to stand the resistor or diode on its end and fold one lead down until it is parallel with the other lead. The component leads can then be placed in virtually adjacent holes.

One final low profile component that is an absolute necessity on any PC board that uses ICs is the IC socket. An IC socket is soldered to the PC board to hold an IC. Therefore, the IC is free to be inserted into or extracted from the socket at any time. Acting as a safety measure, the IC socket prevents any damage that might be caused to a chip by an excessively hot soldering iron, if the IC were soldered directly to the PC board. IC chips are extremely delicate, and both heat and static electricity will damage them. After the project board has been completely soldered, the ICs are finally added.

THE FINISHING TOUCH

The most easily acquired enclosure for your finished project is a metal or plastic cabinet which can be purchased from your local electronics store. Radio Shack makes a stylish, wedge-shaped enclosure (Radio Shack #270-282) and a two-tone cabinet (Radio Shack #270-272 and #270-274) all three of which are perfect for holding your project. Another slightly less attractive, but still useful, project enclosure that is also available from Radio Shack is the Experimenter Box (Radio Shack #270-230 through #270-233 and #270-627). The Experimenter Box comes in five different sizes for holding any size of project.

If none of these enclosures fit your requirements, you can also build your own project cabinet. The best material for building your own enclosure is with one of the numerous, inexpensive types of sheet plastics that are currently available. Materials, such as Plexiglas, are easy to manipulate with the right tools and adhesives. Plexiglas sheeting can be cut with a hand saw or a power jig saw. Just remember not to remove the protective paper covering from the Plexiglas while cutting it. This will prevent the sheet from splitting or becoming scratched.

Construct all sides of the enclosure by adjoining sides together and laying a thin bead of a liquid adhesive along the joint. A powerful cyanoacrylate adhesive, such as Satellite City's SUPER "T" will bond two pieces of Plexiglas together immediately. Be sure that all panel edges are perfectly aligned before applying the adhesive. One side of the final enclosure design should not be joined with the adhesive. This panel will be held on with screws so that it is easily removed for future access to your circuit.

Hopefully, you have finished reading this portion of Appendix A before you have started your project's actual construction. If so, good; you have saved yourself several hours of headaches and, quite probably, several dollars in wasted expenses. If not, all is not lost. Just review your current construction in light of what you have learned in this appendix and make any needed changes in your construction procedures. At least when you make your next project you will know all of the secrets to successful circuit construction.

PCB DESIGN WITH CAD SOFTWARE

For the most part, if you own either an Apple Macintosh or an IBM PC family microcomputer, then you have the potential for designing your own printed circuit boards (PCBs). This potential is only realized after the purchase of some specialized design software, however. A rather general term is applied to this type of software—Computer-aided design or CAD software. CAD is a relatively young field with true dominance already present in the IBM PC arena. This definitive CAD software product is called AutoCAD 2 and it is manufactured by Autodesk, Inc. Don't be fooled by other CAD products that are designed for IBM PCs and their clones and cost less than AutoCAD's roughly $2000 price tag. These cheaper programs are totally inferior to AutoCAD. Without question, the AutoCAD environment is the most powerful CAD software that is currently available for microcomputers and yet it remains flexible to every users' demand.

One fault that is frequently attached to CAD software like AutoCAD is the lack of a dedicated PCB template formulation application. In other words, programs, like AutoCAD, must be customized with user-created symbol libraries before PCB templates can be expertly designed in a minimal amount of time. If you think that there's got to be a better way, then let Wintek Corporation show you the path to this ideal solution. Two professional pieces of CAD software, smARTWORK and HiWIRE, are vertical applications with specific PCB template fabrication virtues. These programs sport an impressive list of PCB template design features:

- ❖ a silkscreen layer
- ❖ text lettering for every layer
- ❖ variable trace width
- ❖ user-definable symbol libraries
- ❖ automatic solder mask generation
- ❖ 2X check prints
- ❖ high-quality dot matrix 2X artwork
- ❖ numerous printer/plotter drivers
- ❖ rapid printouts
- ❖ AutoCAD *.DXF file generation

smARTWORK

In order to provide a more clear picture of the operation of smARTWORK, Figs. A-2 through A-6 illustrate the major steps involved in the design of a PCB. This representative design is the template for the Keyboard Encoder project from Chapter 3.

The bottom line for the Wintek Corporation PCB CAD products is that they will aid you in the design of custom PCB templates for less than half the cost of AutoCAD and with double the creative power.

QUIK CIRCUIT

Until recently, your only PCB CAD route was with the previously described smARTWORK. The cost in the IBM hardware alone was prohibitive to some designers. Most of this need for costly hardware changed with the introduction of the Apple Computer Macintosh. This graphics-based computer seemed ideally suited to the tasks of PCB CAD. Apparently, one different manufacturer shared this same opinion and created a breakthrough product for PCB design. Quik Circuit by Bishop Graphics is a full-featured CAD program that utilizes a graphics environment for preparing PCB layouts.

A major point in the favor of these two PCB design programs is their low cost. Remarkably enough, you could purchase an entire Macintosh CAD system (i.e., 512K Mac, ImageWriter, and either Quik Circuit) for the same cost as AutoCAD 2. Now don't misinterpret this statement as a proclamation of equivalence between this Macintosh CAD package and either AutoCAD 2 or smARTWORK. This just isn't so. The superiority of AutoCAD 2 and smARTWORK over this Macintosh program is clearly definable. This cost

`00.00, 0.00, sol`   `+00.00, +0.00`

Fig. A-2. Designing a PCB template with smARTWORK begins with a blank work area.

COMMAND> dip s 14 3█

Fig. A-3. DIP patterns are placed on the work area with a command line entry.

factor, however, is important to some designers and serves as a yardstick for measuring their solutions. A better and more easily defensed argument that favors all PCB design software is with regard to the time saved over conventional circuit design methods.

ETCHING A PCB

Now that you know how a PCB template is made, your next step is learning how to make a PCB. But what are PCBs anyway? PCBs are vital for the mass production of circuit designs. The printed circuit board serves as the substratum for building any electronics circuit. The board itself is usually constructed from glass epoxy with a coating of copper on one or both of its sides. By using a powerful acid etchant like anhydrous ferric chloride, all of the copper that isn't protected by a resist (a substance that isn't affected by the action of the acid) is eaten away and removed from the PCB. This process leaves behind a copper tracing and/or pad where there should be an electrical connection. The leading method for placing the areas of resist on a PCB is with a photographic negative technique. Kepro Circuit Systems, Inc. markets

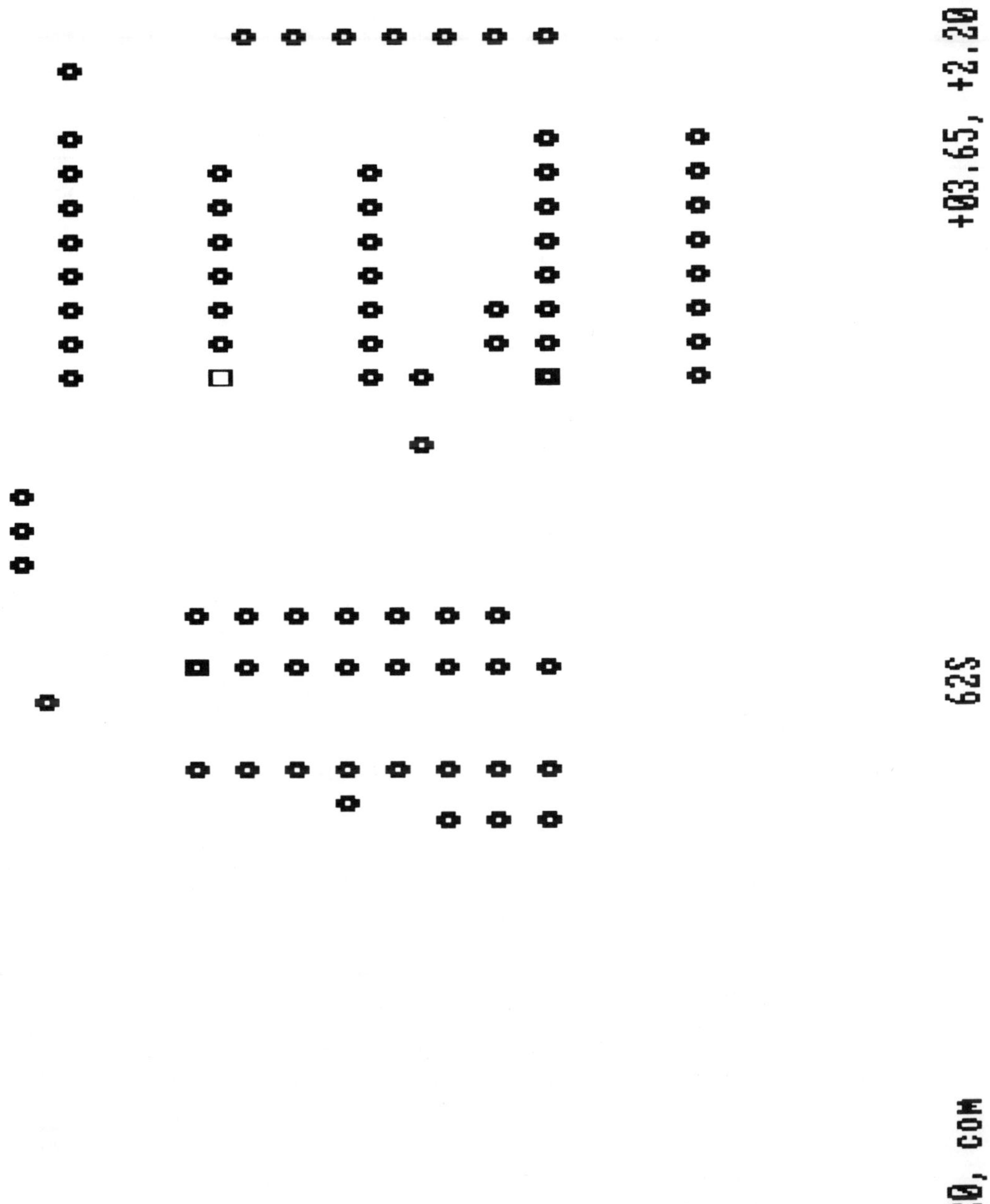

03.65, 2.20, COM   62S   +03.65, +2.20

Fig. A-4. As additional pads are added to the PCB design, the command line features complete location and pad definition information.

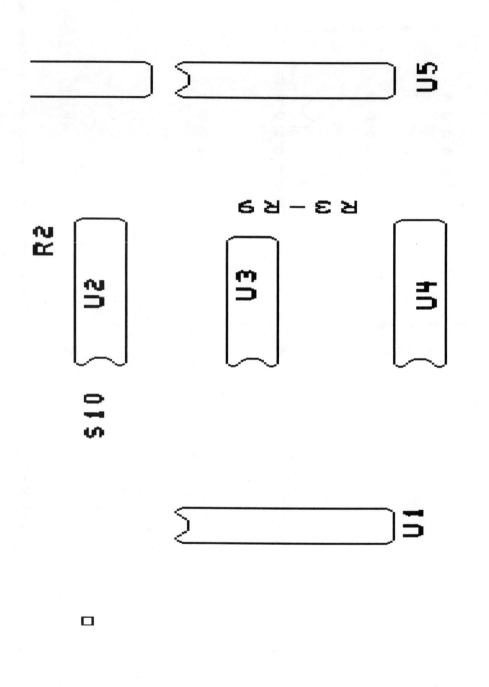

Fig. A-5. A special silkscreen layer provides a detailed visual accounting of the part layout.

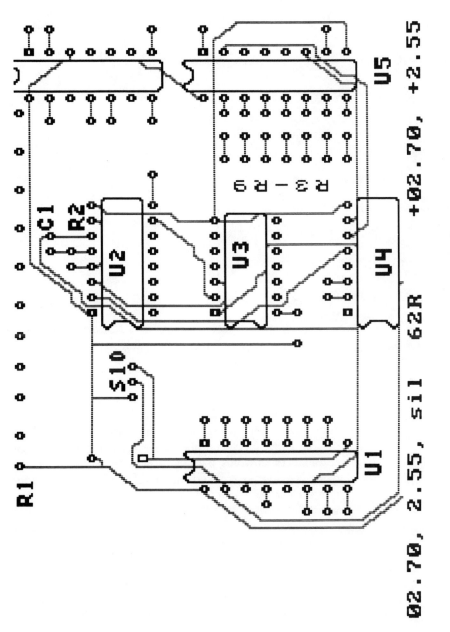

Fig. A-6. An autorouting feature lets smARTWORK perform
all of the steps which are necessary for running traces between selected pads.

Fig. A-7. Perfect PCBs can be created with Kepro materials and the appropriate high contrast negative.

a full range of pre-coated circuit boards for this purpose. These boards are excellent for PCB fabrication, some, like the KeproClad Dry Film KC1-46B, come with their own developer and stripper (see Fig. A-7).

Basically, this technique acts exactly like its paper-based photographic cousin. In other words, parts of the PCB that are protected by the black portions of the negative are covered with a resist, while those regions that are exposed to the clear portions of the negative are removed by the acid.

B

IC Data Sheets

3914

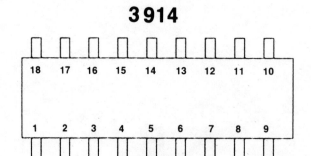

Pin Assignments			
Pin Number	Function	Pin Number	Function
1	LED1	10	LED10
2	V –	11	LED9
3	V +	12	LED8
4	Divider	13	LED7
5	Signal IN	14	LED6
6	Divider	15	LED5
7	Ref OUT	16	LED4
8	Ref Adjust	17	LED3
9	Mode Select	18	LED2

555

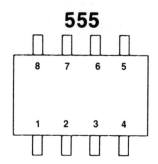

Pin Assignments	
Pin Number	Function
1	GND
2	Trigger
3	Output
4	Reset
5	Control Voltage
6	Threshold
7	Discharge
8	Vcc

7447

Pin Assignments			
Pin Number	Function	Pin Number	Function
1	B IN	9	e OUT
2	C IN	10	d OUT
3	Lamp Test	11	c OUT
4	RB OUT	12	b OUT
5	RB IN	13	a OUT
6	D IN	14	g OUT
7	A IN	15	f OUT
8	GND	16	Vcc

7447

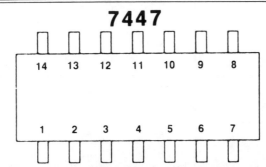

Pin Assignments			
Pin Number	Function	Pin Number	Function
1	1A	8	3Y
2	1B	9	3A
3	1Y	10	3B
4	2A	11	4Y
5	2B	12	4A
6	2Y	13	4B
7	Gnd	14	Vcc

74LS73

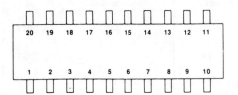

Pin Assignments			
Pin Number	Function	Pin Number	Function
1	Out Control	11	Enable G
2	1Q	12	5Q
3	1D	13	5D
4	2D	14	6D
5	2Q	15	6Q
6	3Q	16	7Q
7	3D	17	7D
8	4D	18	8D
9	4Q	19	8Q
10	Gnd	20	V_{CC}

CT256A-AL2

Pin Assignments			
Pin Number	Function	Pin Number	Function
1	B5/R/W	21	D5
2	B7/CLK OUT	22	D4
3	B0	23	D3
4	B1	24	D2
5	B2	25	V_{CC}
6	A0	26	D1
7	A1	27	D0
8	A2	28	C0
9	A3	29	C1
10	A4	30	C2
11	A7	31	C3
12	INT3	32	C4
13	INT1	33	C5
14	RESET	34	C6
15	A6/SCLK	35	C7
16	A5/RXD	36	MC
17	XTAL2/CLK IN	37	B3/TXD
18	XTAL1	38	B4/Latch
19	D7	39	B6/Enable
20	D6	40	V_{CC}

LM386

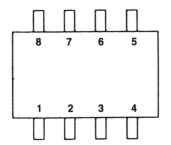

Pin Assignments	
Pin Number	**Function**
1	Gain
2	− Input
3	+ Input
4	Gnd
5	V_{Out}
6	V_S
7	Bypass
8	Gain

MC1488

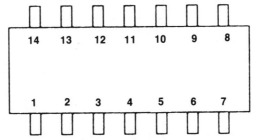

Pin Assignments			
Pin Number	**Function**	**Pin Number**	**Function**
1	V_{EE}	8	Output C
2	Input A	9	Input C_2
3	Output A	10	Input C_1
4	Input B_1	11	Output D
5	Input B_2	12	Input D_2
6	Output B	13	Input D_1
7	Gnd	14	V_{CC}

MC1489

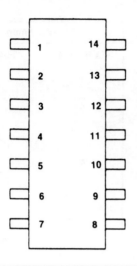

Pin Assignments			
Pin Number	Function	Pin Number	Function
1	Input A	8	Output C
2	Control A	9	Control C
3	Output A	10	Input C
4	Input B	11	Output D
5	Control B	12	Control D
6	Output B	13	Input D
7	Gnd	14	V_{CC}

SPO256-AL2

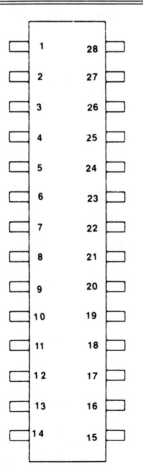

Pin Assignments			
Pin Number	Function	Pin Number	Function
1	V_{SS}	15	A4
2	Reset	16	A3
3	ROM Disable	17	A2
4	C1	18	A1
5	C2	19	SE
6	C3	20	ALD
7	V_{DD}	21	SER IN
8	SBY	22	TEST
9	LRQ	23	V_{D1}
10	A8	24	Audio Out
11	A7	25	SBY RESET
12	SER OUT	26	ROM CLK
13	A6	27	OSC 1
14	A5	28	OSC 2

C

Supply Source Guide

References to a number of unusual materials for constructing CMOS projects have been made throughout this book. Because some of these materials might be difficult to find in many remote areas, this appendix provides a list of mail order houses through which these items can be purchased. Additionally, the names and addresses of specific product manufacturers are included.

Apple Computer, Incorporated
20525 Mariani Avenue
Cupertino, CA 95014
   ImageWriter
   LaserWriter
   Apple II family of computers, including the II+, IIe, and IIc
   Macintosh Computer

Autodesk, Incorporated
2658 Bridgeway
Sausalito, CA 94965
   AutoCAD 2 software

Bishop Graphics, Incorporated
P. O. Box 5007
5388 Sterling Center Drive
Westlake Village, CA 91359
   E-Z Circuit PC Boards (#EZ7402, #EZ7475, and #EZ7464)
   E-Z Circuit Pressure-Sensitive Copper Patterns
   Quik Circuit

Borland International
4585 Scotts Valley Drive
Scotts Valley, CA 95066
   Turbo BASIC
   Turbo C

Bytek Corporation
1021 South Rogers Circle
Boca Raton, FL 33431
   System 125 PROM Programmer
   WRITER-I
   BUV-3 EPROM Eraser

CAD Software, Inc.
P.O. Box 1142
Littleton, MA 01460
   PADS-PCB

Heath/Zenith
Benton Harbor, MI 49022
    ET-3400A Microprocessor Trainer

Intersil, Inc.
10600 Ridgeview Court
Cupertino, CA 95014
    LM2907
    ICM7226B
    ICM7243A
    ICL7107

Jameco Electronics
1355 Shoreway Road
Belmont, CA 94002
    CMOS ICs
    LEDs
    EPROMs
    EPROM Programmer & Eraser

Kepro Circuit Systems, Inc.
630 Axminister Drive
Fenton, MO 63026
    Kepro Pre-sensitized Circuit Boards

Newark Electronics
500 North Pulaski Road
Chicago, IL 60624
    CMOS ICs
    LEDs
    EPROMs

Radio Shack Stores
    64K 2764 EPROM
    3916
    4001
    4011
    4013
    4017
    4049
    4066
    B1001R
    CTS256-AL2
    SPO256-AL2

TLC555
UMC3482
Modular Breadboard Socket

Scott Electronics Supply Corporation
4895 F Street
Omaha, NE 68117
Ungar System 9000
Weller EC2000 Soldering Station

Sig Manufacturing Company, Incorporated
401 South Front Street
Montezuma, IA 50171
Aeroplastic ABS plastic sheeting
Clear Plastic Sheets
X-Acto Saw Blades (#234)

Tower Hobbies
P. O. Box 778
Champaign, IL 61820
Satellite City's Super "T" cyanoacrylate adhesive

VAMP, Inc.
6753 Selma Avenue
Los Angeles, CA 90028
McCAD P.C.B.

Vector Electronics Company, Inc.
12460 Gladstone Avenue
Sylmar, CA 91342
Vector Plugboards

Wahl Clipper Corporation
2902 Locust Street
Sterling, IL 61081
Isotip Soldering Irons (#7800, #7700, and #7240)

Wintek Corporation
1801 South Street
Lafayette, IN 47904-2993
smARTWORK
HiWIRE

Glossary

access time—the delay time interval between the loading of a memory location and the latching of the stored data.

address—the location in memory where a given binary bit or word of information is stored.

allophone—two or more variants of the same phoneme.

alphanumeric—the set of alphabetic, numeric, and punctuation characters used for computer input.

analog/digital (A/D) conversion—a device that measures incoming voltages and outputs a corresponding digital number for each voltage.

ASCII—American Standard Code for Information Interchange.

assembly language—a low level symbolic programming language that comes close to programming a computer in its internal machine language.

binary—the base two number system, in which 1 and 0 represent the ON and OFF states of a circuit.

bit—one binary digit.

byte—a group of eight bits.

CCD—Charge-Coupled Device; a SAM with slow access times.

chip—an integrated circuit.

chip enable—a pin for activating the operations of a chip.

chip select—a pin for selecting the I/O ports of a chip.

CPU—Central Processing Unit; the major operations center of the computer where decisions and calculations are made.

CMOS—a complementary metal oxide semiconductor IC that contains both P-channel and N-channel MOS transistors.

data—information that the computer operates on.

data rate—the amount of data transmitted through a communications line per unit of time.

debug—to remove program errors, or bugs, from a program.

digital—a circuit that has only two states, ON and OFF, which are usually represented by the binary number system.

disk—the magnetic media on which computer programs and data are stored.

DOS—disk operating system; allows the use of general commands to manipulate the data stored on a disk.

EAROM—electrically alterable read only memory; also known as read mostly memory.

EEPROM—electrically erasable programmable read only memory; both read and write operations can be executed in the host circuit.

EPROM—an erasable programmable read-only memory semiconductor that can be user-programmed.

field-programmable logic array—a logical combination of programmable AND/OR gates.

firmware—software instructions permanently stored within a computer using a read only memory (ROM) device.

floppy disk—see disk.

flowchart—a diagram of the various steps to be taken by a computer in running a program.

hardware—the computer and its associated peripherals, as opposed to the software programs that the computer runs.

hexadecimal—a base sixteen number system often used in programming in assembly language.

input—to send data into a computer.

input/output (I/O) devices—peripheral hardware devices that exchange information with a computer.

interface—a device that converts electronic signals to enable communications between two devices; also called a port.

languages—the set of words and commands that are understood by the computer and used in writing a program.

loop—a programming technique that allows a portion of a program to be repeated several times.

LSI—a layered semiconductor fabricated from approximately 10,000 discrete devices.

machine language—the internal, low level language of the computer.

memory—an area within a computer reserved for storing data and programs that the computer can operate on.

microcomputer—a small computer, such as the IBM PC AT, that contains all of the instructions it needs to operate on a few internal integrated circuits.

mnemonic—an abbreviation or word that represents another word or phrase.

MOS—a metal oxide semiconductor containing field-effect MOS transistors.

NMOS—an N-channel metal oxide semiconductor with N-type source and drain diffusions in a P substrate.

Nonvolatile—the ability of a memory to retain its data without a power source.

octal—a base eight number system often used in machine language programming.

opcode—an operation code signifying a particular task to be performed by the computer.

PLA—see field programmable logic array.

parallel port—a data communications channel that sends data out along several wires, so that entire bytes can be transmitted simultaneously, rather than by one single bit at a time.

peripheral—an external device that communicates with a computer, such as a printer, a modem, or a disk drive.

phoneme—the basic speech sound.

PMOS—a P-channel metal oxide semiconductor with P-type source and drain diffusions in an N substrate.

program—a set of instructions for the computer to perform.

RAM—random access memory; integrated circuits within the computer where data and programs can be stored and recalled. Data stored within RAM is lost when the computer's power is turned off.

ROM—read-only memory; integrated circuits that permanently store data or programs. The information contained on a ROM chip cannot be changed and is not lost when the computer's power is turned off.

RS-232C—a standard form for serial computer interfaces.

serial communications—a method of data communication in which bits of information are sent consecutively through one wire.

software—a set of programmed instructions that the computer must execute.

statement—a single computer instruction.

static—a RAM whose data is retained over time without the need for refreshing.

subroutine—a small program routine contained within a larger program.

terminal—an input/output device that uses a keyboard and a video display.

volatile—the inability of a memory to retain its data without a power source.

word—a basic unit of computer memory usually expressed in terms of a byte.

For Further Reading

BOOKS

Mastering the 8088 Microprocessor, 1984, Dao, L.V.
TAB BOOKS, Inc., Blue Ridge Summit, PA
    A thorough examination of the 8088 MPU and its command set

Automatic Translation of English Text to Phonetics by Means of Letter to Sound Rules, 1976, Elovitz, H.S., R.W. Johnson, A. McHugh, and J.E. Shore
United States Naval Research Laboratory Report 7948
    This is the original study that is the basis for most of today's text-to-speech algorithms

Interfacing & Digital Experiments with Your Apple, 1984, Engelsher, C. J.
TAB BOOKS, Inc., Blue Ridge Summit, PA
    All of the elemental electronics that you will need to know for plugging a CMOS circuit into your Apple computer

Using Integrated Circuit Logic Devices, 1984, Horn, D.T.
TAB BOOKS,
    Covers the fundamentals of gates and other digital logic devices

How to Use Special-Purpose ICs, 1986, Horn, D.T.
TAB BOOKS, Inc., Blue Ridge Summit, PA
    An interesting assortment of component data sheets for numerous digital and linear chips

30 Customized Microprocessor Projects, 1986, Horn, D.T.
TAB BOOKS, Inc., Blue Ridge Summit, PA
    30 different Z80-based projects are described along with an EPROM programmer

101 Projects, Plans, and Ideas for the High-Tech Household, 1986, Knott, J. and D. Prochnow
TAB BOOKS, Inc., Blue Ridge Summit, PA
    Over one hundred circuit designs and ideas; many of which use CMOS technology

Microprocessors and Logic Design, 1980, Krutz, R. L.
John Wiley & Sons, New York, NY
    A functional introduction into MPU and memory interfacing theory

Microprocessor Architecture and Programming, 1977, Leahy, W. F.
John Wiley & Sons, New York, NY
    A beginning text on implementing digital microcomputer designs

The Handbook of Microcomputer Interfacing, 1983, Leibson, S.
TAB BOOKS, Inc., Blue Ridge Summit, PA
    An excellent introduction to the electronics of parallel and serial connections

Troubleshooting and Repairing the New Personal Computers, 1987, Margolis, A.
TAB BOOKS, Inc., Blue Ridge Summit, PA
    This book's odd title doesn't adequately convey the wealth of information on general microcomputer circuit design that is contained inside

Chip Talk: Projects in Speech Synthesis, 1987, Prochnow, D.
TAB BOOKS, Inc., Blue Ridge Summit, PA
    The definitive source on digital speech synthesis theory and speech synthesizer design

Digital Processing of Speech Signals, 1978, Rabiner, L.R. and R.W. Schafer
Prentice-Hall, Inc., Englewood Cliffs, NJ
    A technical look at the formulas of the different speech synthesis techniques

Mastering the 68000 Microprocessor, 1985, Robinson, P.R.
TAB BOOKS, Inc., Blue Ridge Summit, PA
    Complete data on the structure and command set found in the 68000 family of MPUs

Microprocessors and Programmed Logic, 1981, Short, K. L.
Prentice-Hall, Inc., Englewood Cliffs, NJ
    All of the theory of MPU and memory interfacing that you'll ever need

101 Projects for the Z80, 1983, Tedeschi, F. P. and R. Colon
TAB BOOKS, Inc., Blue Ridge Summit, PA
    Hardware and software projects geared for the SD-Z80 System

MAGAZINE ARTICLES

ed., "Monolithic Memories Goes to CMOS for Its PALs," *Electronics*, August 6, 1987

Cohen, C. L., "NEC's BiCMOS Arrays Shatter Record,"*Electronics*, August 6, 1987

Cole, B. C., "Alliance's 1-MBit DRAM Runs Fast as a Static RAM," *Electronics*, January 7, 1988

Cole, B. C., "Is BiCMOS the Next Technology Driver?," *Electronics*, February 4, 1988

Lineback, R. L., "CMOS A-to-D Converter Runs Twice as Fast," *Electronics*, November 26, 1987

Weber, S., "TI Soups Up LinCMOS Process with 20-V Bipolar Transistors," *Electronics*, February 4, 1988